PRINCIPLES OF STABLE ISOTOPE DISTRIBUTION

PRINCIPLES of STABLE ISOTOPE DISTRIBUTION

Robert E. Criss

New York   Oxford

Oxford University Press

1999

Oxford University Press

Oxford   New York
Athens   Auckland   Bangkok   Bogotá   Buenos Aires   Calcutta
Cape Town   Chennai   Dar es Salaam   Delhi   Florence   Hong Kong   Istanbul
Karachi   Kuala Lumpur   Madrid   Melbourne   Mexico City   Mumbai
Nairobi   Paris   São Paulo   Singapore   Taipei   Tokyo   Toronto   Warsaw

and associated companies in
Berlin   Ibadan

Published by Oxford University Press, Inc.
198 Madison Avenue, New York, New York 10016

Oxford is a registered trademark of Oxford University Press

Library of Congress Cataloging-in-Publication Data
Criss, R. E.
    Principles of stable isotope distribution / by Robert E. Criss.
        p.   cm.
    Includes bibliographical references and index.
    ISBN -13 978 - 0-19-511775 -2
    ISBN 0-19-511775-1
    1. Stable isotopes.   I. Title.
QD466.5.C1C75   1999
541.3'88—dc21   98-24609

9 8 7 6

Printed in the United States of America
on acid-free paper

Preface

The field of stable isotope distribution is a product of the spectacular advances in physics and chemistry that were realized in the first half of the twentieth century. The new model for the atom led to the discovery of the neutron, to H. C. Urey's discovery of deuterium, and ultimately to Urey's quantitative prediction of the isotope fractionation effects that form the cornerstone of this field. Accompanying these advances was the development by A. O. Nier of mass spectrometers that could measure the relative abundances of stable isotopes to high levels of precision.

The subsequent 50 years witnessed the blossoming of this knowledge into many new applications and subfields. Stable isotope data are now frequently used in the fields of hydrology, cosmochemistery, geochemistry, igneous petrology, paleoclimatology, glaciology, economic geology, and oceanography, and are seeing increased use in atmospheric and planetary science, microbiology, agronomy, physiology, contaminant transport, ecology, and forensics. New instrumental designs have made many types of measurements rapid and routine, while other developments have pushed analytical capability to higher levels of precision on isotopes with lower abundance levels in smaller and smaller samples.

From the perspective of the late 1990s, it is interesting to ask which part of the century, the first or the second, produced the most important results. Clearly, the laurels must go to the former, as it is during this time that the fundamental basis of the field was established. In contrast, the second half of the century has mostly embellished the details of isotope distribution, and while many of the associated discoveries have considerable scientific and practical utility, none have the fundamental importance of several of the earlier developments. When it is realized that the expenditure of funding and manpower in this area in the second half of the century has probably been at least $100\times$ greater than in the first, what at first may appear to represent glorious postwar progress might more correctly be viewed as significant underachievement.

The reason for this disparity in accomplishment is obvious. This difference is the product of too little attention to fundamentals, too much emphasis on specialized details, too much relaxation of academic standards, and misdirection of federal support. Several academic posts in this field are now held by individuals who have little competence in physical chemistry or differential equations, and even by some who have never had a single course in either subject. After all, these courses are now no longer required for the baccalaureate science degrees offered by many departments at numerous universities. Even the National Science Foundation (NSF) has steadily redirected its support from fundamental science to topics, such as climatic change, that it views as politically relevant. As a consequence, several of the finest research groups in this field have struggled for years with little or no federal support, while several of the feeblest have been abundantly nurtured.

This book represents my reaction to this personal view of these trends and times. Specifically, the book seeks to reconnect the diverse observations of isotope distribution to the quantitative theories of physical chemistry emphasized by earlier scientists. Case histories are de-emphasized, being presented only when they exemplify quantitative principles or convey new and important possibilities. In their place are translations of the principles of statistical and classical thermodynamics, kinetics, and diffusion theory into the language of isotope distribution, whose fundamental variable is the isotope ratio R. A very large number of the equations are new, but this fraction increases from only a little in the beginning of the book to nearly half in the final chapters. Many important differential equations are corrected and generalized, in some cases to include isotopes that are not present in trace concentrations. Numerous exact solutions to these equations are provided for different cases of interest. The emphasis on the isotope ratio R, rather than on the delta-value δ, will doubtless trouble many readers accustomed to the conventional notation, but the delta-notation obfuscates physical principles, needlessly complicates mathematical relationships, and had led to numerous unnecessary approximations in previous literature that have impeded theoretical progress. Appendix A.3 provides several identities that may be used to easily translate various terms that involve isotope ratios into terms that represent delta-values.

The book is organized into five chapters, each followed by suggested quantitative problems and a short reference list. Key equations are "boxed" for emphasis; the important ones are referred to as "(R. E. C., unpublished)" if it is believed that they are new or significantly corrected. This procedure is not a matter of gratuitous self-reference but rather is intended to emphasize new quantitative possibilities and to identify results that need confirmation. My apologies are extended in advance to any individuals unintentionally slighted by rederivation of their published work. Most of the equations have diverse applications to many isotopic systems, but the specific examples provided here primarily relate to the distribution of hydrogen and oxygen isotopes, as these are well studied, exhibit large effects, and represent abundant and ubiquitous elements. The problems are mostly of moderate difficulty and are designed to illustrate the principles and equations of the chapter, sometimes building on earlier chapters. The appendices provide data on atomic weights of light nuclides, physical constants, mathematical relationships, and isotopic fractionation factors. As such, the book should serve as a textbook for advanced students, as a research reference, and in some cases as a quick source of information.

Chapter 1 is an elementary review of atomic structure, the distribution and fractionation of isotopes, and the physics of mass spectrometers. Chapter 2 links isotope distribution to thermodynamic relationships, reviews Urey's theory of isotopic fractionation, and discusses the rule of the mean and isotope thermometry. New relationships are presented for isotopic fractionations for elements that have more than two stable isotopes. Chapter 3 shows how thermodynamics and fractionation effects control the distribution of hydrogen and oxygen isotopes in natural waters and the atmosphere. The differential equation for a Rayleigh process is extended and generalized, and new and highly successful models are presented for the hydrograph and for the isotopic variations in streamflow. Chapter 4 discusses isotopic exchange and isotope distribution in disequilibrium systems. New equations extend the author's kinetic theory of isotopic exchange to many new cases, and several testable predictions are made along the way. The chapter includes a discussion of evaporation, shows how diffusion equations may be used to predict spatial variations in disequilibrium systems, and concludes with atmospheric processes, such as mass-independent fractionation, gravitational fractionation, and loss to space. Chapter 5 discusses isotopic distribution in the solid Earth and in extraterrestrial materials, with an emphasis on igneous processes and fluid–rock interactions that exemplify the principles of mass balance and isotopic exchange. Throughout the book, emphasis is placed on eliminating unconstrained parameters from the equations, so that testable formulations that correlate measurable system variables are the result.

The recent production of this book belies its origins, more than 25 years ago, when Samuel M. Savin introduced me to stable isotope geochemistry in his class, and opened my eyes to an exciting career. Between then and now, my intellectual debts to colleagues, collaborators, and mentors have steadily grown. While inclusion in this list does little justice to their contributions, the following individuals have all significantly furthered my understanding of isotope distribution: J. G. Arth, D. E. Champion, C. B. Douthitt, S. Epstein, R. J. Fleck, R. T. Gregory, A. M. Hofmeister, N. L. Ingraham, R. J. Kamilli, M. A. Lanphere, J. R. O'Neil, W. C. Shanks, H. P. Taylor, Jr., and C. J. Yapp. However, my debt to my former students M. L. Davisson, G. F. Eaton, G. C. Frederickson, E. B. Melchiorre, and T. P. Rose exceeds even this. The course lectures that form the basis for this book were inspired and augmented largely in response to their interest, loyalty, dedicated and careful work, penetrating questions, and their acquisition of important data sets that demanded explanation. It gives me the greatest personal pleasure to have seen each of these individuals develop research capabilities in various areas that substantially exceed my own.

Saint Louis   Robert E. Criss
March 1998

Contents

PRINCIPLES OF
STABLE ISOTOPE
DISTRIBUTION

I

Abundance and Measurement of Stable Isotopes

1.1 Discovery of Isotopes

The discovery of isotopes is best understood in the context of the spectacular advances in physics and chemistry that transpired during the last 200 years. Around the year 1800, compounds and elements had been distinguished. About 39 elements were recognized, and discoveries of new elements were occurring rapidly.

At about this time, the chemist John Dalton revived the ancient idea of the atom, a word derived from the Greek "atomos," which literally means "indivisible." According to Dalton's theory, all matter is made of atoms which are immutable and which cannot be further subdivided. Moreover, Dalton argued that all atoms of a given element are identical in all respects, including mass, but that atoms of different elements have different masses.

Even today, Dalton's atomic theory would be accepted by a casual reader, yet later developments have shown that it is erroneous in almost every one of its key aspects. Nevertheless, Dalton's concept of the atom was a great advance, and, with it, he not only produced the first table of atomic weights, but also generated the concept that compounds comprise elements combined in definite proportions. His theory laid the groundwork for many other important advances in early nineteenth-century chemistry, including Avogadro's 1811 hypothesis that equal volumes of gas contain equal numbers of particles, and Prout's 1815 hypothesis that the atomic weights of the elements are integral multiples of the weight of hydrogen.

By 1870, approximately 65 elements had been identified. In that year, Mendeleev codified much of the available chemical knowledge in his "periodic table," which basically portrayed the relationships between the chemical properties of the elements and their atomic weights. The regularities that Mendeleev found directly lead to the discovery of several "new" elements—for example, Sc, Ga, Ge, and Hf—that filled vacancies in his table and confirmed his predictions of their chemical properties and atomic weights. Similarly, shortly after Rayleigh and Ramsay isolated Ar from air in 1894, the element He was isolated from uranium minerals in 1895; the elements Ne, Kr, and Xe were found in air in 1898; and Rn was discovered in 1900. In other words, as soon as it became apparent that an entire column of the periodic table had been missed, enough was known about the properties of the "missing" elements to allow for the rapid discovery of the entire group.

By the turn of the century, the periodic table was essentially complete up to the element U. However, a number of new problems appeared that could not be explained. The first relates to the atomic masses themselves. Careful measurements had shown that the atomic weights of the elements did not always increase with the atomic number and therefore with the position in the periodic table. In fact, one can readily confirm that the atomic weight of Ar exceeds that of the succeeding element, K. Similarly, the atomic weight of Co exceeds that of Ni, and the atomic weight of Te exceeds that of I. In addition, T. W. Richards established that the atomic weights of elements are clearly *not* simple integral multiples of the mass of hydrogen, contrary to Prout's hypothesis. Even more remarkable was Richards' meticulous demonstration that samples of lead from different localities had different atomic weights!

An entirely new class of problems arose regarding the indivisibility of the atom. Most important was the discovery of radiation, an entity that cannot be seen, heard, or felt. This advance included the discovery of X-rays by Roentgen in 1895, the discovery of "cathode rays" by J. J. Thomson in 1897, and the discovery of the spontaneous emission of radiation by uranium salts by Becquerel in 1896. Adding to the confusion was the subsequent discovery that different kinds of Th had different rates of decay.

In 1914, Frederick Soddy proposed a resolution to the above enigmas by arguing that the place occupied by a particular element the periodic table can accommodate more than one kind of atom. Soddy coined the term "isotopes," literally meaning "same place," for these atoms. A standard definition for isotopes would be

Isotopes ≡ Nuclides of a single element that have different atomic weights

where the word "nuclide" refers to any distinctive type of atom. Note that Soddy's concept differs significantly from Dalton's. Specifically, the identical nature of Dalton's atoms is incompatible with the different masses and different decay rates now known to characterize the various isotopic forms of certain elements. Further, the indivisibility and immutability of Dalton's atoms contrast with the radioactive emissions of certain elements, be they spontaneous or induced.

At the present time, more than 2500 different nuclides are known, yet together these constitute only about 110 different chemical elements. While most nuclides are radioactive, half of the elements have at least two stable nuclides. The abundance patterns and atomic weights of the nuclides hold the key to the origin of matter; the age of the universe; the source of solar energy; and the origin, age, and differentiation of the Earth.

1.1.1 Confirmation of Isotopes

Thomson's experiments with cathode rays also demonstrated that matter has electrical properties. Utilizing a modified "cathode ray tube," which in effect was the prototype of the mass spectrometer, Thomson was able to measure the charge-to-mass ratio, q/m, of the cathode rays, now called electrons. Thomson demonstrated that this quantity was independent of the type of gas used in the cathode ray tube, and thereby proved that electrons are uniform, corpuscular fragments of atoms (Table 1.1).

Thomson modified his apparatus in order to determine the q/m ratio of heavier particles. In 1913, he was able to show that the element neon comprises two different types of atoms with masses of 20 and 22. This result fully confirmed the existence of isotopes. F. W. Aston further improved Thomson's apparatus by constructing a device he called a "mass spectrograph." This design was so useful that by 1919 he had discovered 212 of the 287 naturally occurring isotopes (Faure, 1986).

1.1.2 A New Model for the Atom

One more key piece of evidence was needed to derive a more realistic model of the atom. This was provided in 1911 by the famous scattering experiment of Ernest Rutherford. Basically, when Rutherford shot energetic α-particles (nuclei of He atoms) at thin metal foils, he found that most of the particles passed through the foils undeflected, but that a few α-particles were deflected through angles as large as 180°. From these experiments and the results of Thomson, Rutherford concluded that atoms have a very small (ca. 10^{-14} m) but massive, positively charged

Table 1.1 Important Nuclear Entities and Emanations

Particle	Symbol	Rest Mass (amu)	Charge
Proton	p	1.0072765	+1
Neutron	n	1.0086649	0
Electron	β^-	0.0005486	−1
Positron	β^+	0.0005486	+1
Gamma-ray	γ	0	0
Protium atom	^{1}H	1.007825	0
Deuterium atom	^{2}H or D	2.014102	0
Alpha-particle	^{4}He^{2+} or α	4.001475	+2

After Walker et al., 1989.

nucleus, surrounded by electrons that orbit in a roughly spherical volume with a radius of about 10^{-10} m (Figure 1.1).

This new concept was very useful, yet no explanation was provided for the existence of isotopes. It remained for Chadwick to discover another fundamental particle, termed the neutron, in 1932. This particle had been produced in an experiment in which a foil of Be metal was bombarded by α-particles. One of the products was a particle that had a mass almost identical to that of the proton, yet had no charge (Table 1.1). This neutral particle provided the explanation for the existence of isotopes, because atoms that have different numbers of neutrons could still have the same nuclear charge. Such atoms could therefore have different masses while exhibiting the chemical similarities common to the atoms of a single element.

These discoveries led to a greatly improved model for the atom in terms of its principal constituents, namely the positively charged protons, the negatively charged electrons, and the uncharged neutrons. More than 99.9% of the mass of the atom, all of the protons and neutrons, and all of the positive charge, reside in the minuscule nucleus that constitutes only about one-trillionth of the atomic volume (Figure 1.1). In contrast, virtually all of the volume of the atom, and all of the electrons that carry all the negative charge, reside in the surrounding electronic shells that contribute almost negligibly to the atomic mass. The chemical behavior of atoms principally relates to the configuration and occupancy of the outermost, chemically interactive electronic shells.

1.1.3 The Mass Number

The concept outlined in section 1.1.2 is the familiar "planetary" model for the atom, taught in elementary chemistry courses. In terms of this model, the basic building blocks of the atom can be accounted for as follows (Table 1.1). The number of protons in the atom, which corresponds to the atomic number and the position of the particular element in the periodic table, is denoted by the symbol Z. In a neutral atom, the number of electrons must also be Z. The number

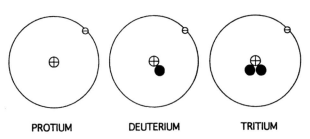

PROTIUM   DEUTERIUM   TRITIUM

Figure 1.1 Diagram of the isotopes of hydrogen, notably protium (^{1}H), deuterium (^{2}H), and tritium (^{3}H), the latter of which is radioactive. The protons (circle with cross), neutrons (filled circle), and electrons (small circle with dash) are shown. All hydrogen atoms have a single proton, but the number of neutrons may vary. The diagram is not to scale, as the Bohr radius of the hydrogen atom is 0.529×10^{-10} m, or 0.529 Å, while the radius of a hydrogen nucleus is only about 10^{-15} m.

of neutrons, which accompany the protons in the nucleus, is denoted by $\mathcal{N}$. Since the protons and neutrons have nearly equal mass, and are much more massive than the electrons, then the mass number A of an atom is defined by the sum

$$\boxed{A = Z + \mathcal{N}} \tag{1.1}$$

where A, Z, and $\mathcal{N}$ are all integers. Because the masses of the protons and neutrons are both rather close to 1 amu (Table 1.1), the atomic weight of a given nuclide, though not an integer, is numerically close to the integral value of A.

A standard notation has been devised to represent different nuclides in terms of these basic constituents. Key to this notation is the ordinary chemical symbol "Γ," preceded by a superscript that indicates the mass number A; that is,

Notation: $^A\Gamma$

For example, the important isotope "carbon-13" is succinctly denoted by ^{13}C. The symbol C automatically conveys the information that the atomic number Z is 6, since carbon is the sixth element in the periodic table. Since the indicated mass number is 13, the number of neutrons is understood to be 7 by difference (equation 1.1).

As a rule of thumb, the numbers of protons and neurons are similar for the stable nuclides of light elements (Figure 1.2); that is, $Z \sim \mathcal{N}$. However, neutrons become more abundant relative to protons in the nuclei of progressively heavier elements. The neutrons seemingly act to overcome the repulsive forces between the protons in the tightly packed atomic nuclei. While no such repulsive forces can exist in protium (^{1}H), where no neutron is needed, such forces must become progressively larger as the atomic number increases.

1.1.4 Moseley's Law

While an elementary model of the atom, sufficient for present purposes, is provided above, not all matters have been accounted for. One fundamental aspect, little remembered nowadays, is how we know that the elements in the periodic table are listed in their proper order. It has been pointed out that the atomic weights of the elements generally, but not perfectly, increase with increasing atomic number. We could exploit the well-known regularities in the periodic table to assign Ar, an inert gas, and K, an alkali metal, to their proper places, in spite of any uncertainty that results from the reversal in their atomic weights. Plausible chemical arguments could also be used to correctly order Te and I. However, Co and Ni are both group VIIIB elements, both have valences of +2 and +3, and both have identical Pauling electonegativities (1.8), so the reversal in their atomic weights is more problematical.

It is useful here to recall the atomic model developed by Niels Bohr in 1913. Bohr combined aspects of Max Planck's quantum theory with Rutherford's "planetary" model of the atom, and developed his well-known theory that retains

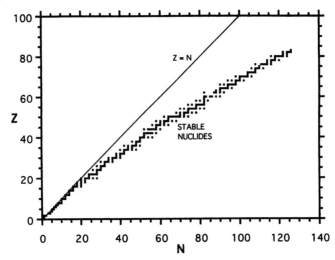

Figure 1.2 Plot of atomic number Z versus the neutron number $\mathcal{N}$ for the 264 stable nuclides. Elements with low atomic numbers have approximately equal numbers of protons and neutrons, while an excess of neutrons is needed to stabilize the nuclei of heavier elements. Nuclei having too many neutrons to be stable tend to emit β^--particles, while those with too few neutrons are commonly positron (β^+) emitters. ^{209}Bi is the heaviest stable nuclide, and plots at the upper right of the diagram. All nuclides with larger mass numbers are radioactive, and many of them are α-emitters.

enough of a classical character to be easily visualized. In effect, Bohr's theory is based on the coulombic interactions of the protons in the nucleus with electrons that move in quantized, nonrelativistic, circular orbits. The biggest success of his theory is that it reproduces and explains an empirical spectral formula developed by J. J. Balmer in 1885. In particular, it provides a theoretical and quantitative description for the emission and absorption spectra of "hydrogen-like" atoms, a group that includes ions such as He$^+$ that have only one electron. Specifically,

$$\frac{1}{\lambda} = \mathcal{R}Z^2\left\{\frac{1}{n_1^2} - \frac{1}{n_u^2}\right\} \tag{1.2}$$

Here, λ is the wavelength of the emission line, $\mathcal{R}$ is the Rydberg constant which is numerically equal to 1.0974×10^7/m, and n_1 and n_u are integers that respectively refer to the lower and upper electronic shells involved in a particular transition.

Due to the complexity of multielectron atoms, no similarly successful model can be derived for heavier elements. However, there is a special and important case of X-ray emissions that can be usefully treated with a simple extension of Bohr's model. X-rays are emitted when a metallic target is bombarded with high-energy electrons, wherein the metal emits a series of spectral lines that are characteristic of the particular element. The systematic study by H. E. J. Moseley in 1913 revealed that the characteristic X-ray spectra of the elements decrease in exactly the same sequence as their order in the periodic table.

Bohr's model can be easily extended to a special case of the X-ray lines, notably the intense K_α lines that are important in X-ray diffractometry. These lines represent the electronic transitions from the second electronic shell to the innermost shell, following the ejection of one of the innermost electrons by external excitation. For atoms in this latter excited configuration, the innermost shell has only one, spherically symmetric electron which shields the nucleus, so the effective coulombic force on the second-shell electrons is therefore approximately $Z - 1$. This result obtains because any electrons in higher shells have little effect on interior electrons, somewhat analogous to the familiar situation in electrostatics where a point charge located anywhere inside a charged, spherical, conducting metal shell experiences no net coulombic force. Thus, one can derive the K_α spectra by a simple modification of equation 1.2, specifically by substituting $Z - 1$ for the nuclear charge Z, and by incorporating the appropriate values of 1 and 2 for n_1 and n_u:

$$\frac{1}{\lambda_{K_\alpha}} = \mathscr{R}(Z - 1)^2 \left\{ \frac{1}{1^2} - \frac{1}{2^2} \right\} \tag{1.3a}$$

This may be inverted to give

$$\lambda_{K_\alpha} = \frac{4}{3\mathscr{R}(Z - 1)^2} \tag{1.3b}$$

On substitution of the numerical constants, equation 1.3b becomes

$$\lambda_{K_\alpha} = \frac{1215.0}{(Z - 1)^2} \tag{1.3c}$$

where λ is in angstroms (Å).

It is useful to compare this simple model with observation. The wavelength of the experimentally observed $K_{\alpha 1}$ lines, $\lambda_{K_{\alpha 1}}$, is plotted versus $1/(Z - 1)^2$ in Figure 1.3. The least-squares regression of the data is

$$\lambda_{K_\alpha} = \frac{1195.3}{(Z - 1)^2} + 0.0198 \tag{1.4}$$

Note that this empirical relationship is almost perfectly linear, having a correlation coefficient of 0.99996, and passes almost perfectly through the origin. Moreover, the observed slope is within 2% of the value given by the simple model represented by equation 1.3c.

Moseley's measurements of X-ray lines clearly established the values of the atomic numbers of the elements. His results showed that the number of extranuclear electrons was identical to the atomic number, and that the atomic number was equal to the nuclear charge. In other words, the atomic number is a fundamental quantity, but the atomic weight of an element is not!

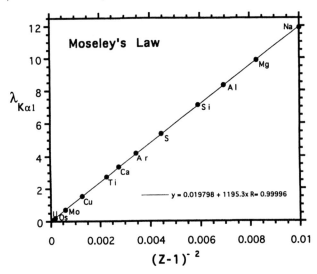

Figure 1.3 Graph of the observed wavelengths (λ, in angstroms) of $K_{\alpha1}$ radiation for several elements (points, data of Lide, 1991) plotted against the squared reciprocal of the atomic number minus 1. The line and the regression are the same as equation 1.4.

1.2 Nuclide Types, Abundances, and Atomic Weights

1.2.1 Types of Nuclides

The > 2500 known nuclides may be characterized as either radioactive or stable. The latter may be subdivided into the radiogenic and the nonradiogenic stable nuclides, depending on their origin.

1.2.1.1 Radioactive Nuclides   Radioactive nuclides decay into different forms, called "daughter" atoms, at statistically predictable rates. These nuclear transformations are typically accompanied by the emanation of radiation such as α-particles ($^4He^{2+}$), β-particles (electrons or positrons), or γ-rays (electromagnetic radiation; see Table 1.1). Rutherford and Soddy found in 1902 that the "activity," representing the number of decay events per unit time, was equal to the number of radioactive atoms present, multiplied by a proportionality constant that is a characteristic of the particular nuclide. The differential equation describing this process is therefore

$$dN/dt = -\lambda N \qquad (1.5)$$

where N is the number of radioactive atoms present at any time, and dN/dt represents the activity. The decay constant, traditionally represented by the symbol λ, is not to be confused with the wavelength of electromagnetic radiation used in the previous section. Here, λ has the physical interpretation of being the reciprocal of the mean life τ of the radioactive atoms. Equation 1.5 may be directly integrated to give the familiar law of radioactive decay:

$$N = N_i e^{-\lambda t}$$
(1.6)

where N_i is the number present at some initial time. Equation 1.6 may be used to show that the "half-life," representing the time interval required for one-half of the atoms of a parent nuclide to decay (i.e., $N = 0.5\,N_i$), is numerically equal to the quantity, $(\ln 2)/\lambda$.

Most of the 2500 known nuclides are artificial radioisotopes with half-lives of seconds or less. In contrast, the nuclides ^{40}K, ^{87}Rb, ^{235}U, ^{238}U, and ^{232}Th have half-lives of 0.7–48 billion years and are the principal contributors to Earth's heat generation. Even these enormous half-lives are dwarfed by those of certain isotopes. For example, Bernatowicz et al. (1992) demonstrated the presence of daughter ^{128}Xe in telluride minerals that was produced by the decay of ^{128}Te with a calculated half-life of 7.7×10^{24} years!

1.2.1.2 Stable Nuclides   Stable nuclides do not decay, although it is possible that some have half-lives that are so long that their radioactivity has not been detected. The *radiogenic stable nuclides* are continuously formed by the decay of a radioactive parent nuclide. Examples of these are ^{40}Ar, ^{87}Sr, and ^{207}Pb. Variations in their abundance are primarily caused by the amount of parent nuclides present, and the amount of time, although processes such as diffusion can be important controls. It is the process of radioactive decay in environments that have different Th/U ratios that gave rise to the observation by Richards that different samples of lead have different atomic weights, because the ultimate daughter product of ^{232}Th decay is ^{208}Pb, whereas that of ^{235}U is ^{207}Pb, and that of ^{238}U is ^{206}Pb.

Because the radioactive atoms do not "disappear," but instead change into another atomic form, in a closed system the sum of the surviving parent atoms (N) plus the daughter (D^*) atoms produced is constant and equal to N_i. Equation 1.6 may be directly rewritten in a form that allows the measurable, present-day concentrations of parent and radiogenic daughter atoms to be related to the elapsed time:

$$D^* = N(e^{\lambda t} - 1)$$
(1.7)

The parent–daughter relationships of certain nuclide pairs provide the basis of the field of geochronology, and are discussed at length in numerous books (e.g., Faure, 1986). Such relationships provide the key to establishing the age of terrestrial and extraterrestrial materials, including the age of the Earth.

Nonradiogenic stable nuclides are the primary focus of this book. These atoms have all been present since the "Big Bang," so their overall abundance is not a function of time. Important examples are ^{13}C, ^{12}C, and ^{18}O. Variations in the abundance patterns of these nuclides may be caused by certain physiochemical processes, as discussed below. In addition to their varied uses as materials, their utility to science lies in their use in isotopic tracer studies and as monitors of physical processes and past conditions.

1.2.2 Isotopic Abundances and Atomic Weights:

Every element consists of an assemblage of different isotopes that usually, but not always (e.g., F, Na, Al, P), incorporate two or more distinct stable isotopes. While nuclides of a given element all share the same atomic number, representing the nuclear charge and the number of protons, the number of neutrons will vary, and so must the mass number A.

The atomic weights and abundances of the nuclides are generally determined by mass spectrometers. The peak position recorded by the instrument relates to the atomic mass, while the abundances are given by the relative peak heights of the separated mass beams. Table 1.2 and Appendix A.1 report the atomic weights of the stable nuclides of several light elements in atomic mass units (amu), where 1 amu is defined as 1/12 the mass of an atom of ^{12}C. The relative abundances are given in atom %.

The atomic weights of the elements (bold numbers in Table 1.2) correspond to the quantities listed in an ordinary periodic table. These weights are not fundamental physical quantities but represent the sum of the masses of the constituent isotopes weighted by their abundances. In contrast, the atomic weights of the individual nuclides are intrinsic and invariant.

Thus, the atomic weight of any given element simply represents the weighted average of the atomic weights of the constituent isotopes, calculated as follows:

$$\boxed{Atomic\ wt. = \sum_j Ab_j Wt_j} \qquad (1.8a)$$

where

$$\sum_j Ab_j = 1 \qquad (1.8b)$$

Table 1.2 Atomic Weights and Abundances of the Stable H, C, N, and O Isotopes

Element	Isotope	Atomic Weight (amu)	Abundance (atom %)
Hydrogen ($Z = 1$)		**1.0079**	
	^1H (Protium)	1.007825	99.985
	^2H (D or Deuterium)	2.014102	0.015
Carbon ($Z = 6$)		**12.011**	
	^{12}C	12.00000	98.90
	^{13}C	13.00335	1.10
Nitrogen ($Z = 7$)		**14.0067**	
	^{14}N	14.003074	99.63
	^{15}N	15.000109	0.37
Oxygen ($Z = 8$)		**15.9994**	
	^{16}O	15.994915	99.76
	^{17}O	16.999131	0.04
	^{18}O	17.999160	0.20

Source: Walker et al., 1989.

where Ab_j and Wt_j refer to the relative abundance and intrinsic atomic weight of the jth nuclide. The element chlorine provides a fine example. Chlorine comprises two stable isotopes, ^{35}Cl and ^{37}Cl, which respectively constitute 75.77% and 24.23% of the atoms. These nuclides have respective atomic weights of 34.96885 and 36.96590 amu (Appendix A.1). The atomic weight of chlorine may then be directly calculated as follows:

$$\text{Atomic wt.} = 0.7577(34.96885) + 0.2423(36.96590) = 35.453 \text{ amu} \qquad (1.8c)$$

This calculation is in agreement with the value reported in a standard periodic table. Note that this average is neither close to an integral multiple of the mass of hydrogen, nor to an integral multiple of the amu. Neither is this mass close to the intrinsic atomic weight of any known nuclide. Moreover, were a sample obtained in which ^{35}Cl and ^{37}Cl were present in proportions different than those given above, then the atomic weight of that sample of chlorine would be different than 35.453!

1.2.3 The Even–Odd Effect

Properties of the approximately 2500 known nuclides have been tabulated by Lide (1991) and Walker et al. (1989). Only about 287 nuclides are either stable or very long-lived radioactive types, but the abundance patterns among these naturally occurring nuclides are of particular interest.

An important observation, known as "Harkin's rule," is that elements having even atomic numbers—that is, paired protons—are significantly (ca. 10×) more abundant in nature than adjacent, odd-numbered elements in the periodic table (e.g., Suess, 1987). Protium (^1H) is an exception, and a noteworthy one because it is the most abundant nuclide in the universe, but this is probably a unique case because there are no proton–proton repulsions in its nucleus. Two odd-numbered elements, technetium ($Z = 43$) and promethium ($Z = 61$), do not even occur naturally on Earth, because all isotopes of these elements have rather short half-lives. Also, no elements with atomic numbers greater than bismuth ($Z = 83$) are stable, although some even-numbered, heavy elements, such as thorium ($Z = 90$) and uranium ($Z = 92$), have some very long-lived isotopes that occur in Earth's crust.

The data in Table 1.2 and Appendix A.1 show that similar effects also apply to the isotopes of a single element, in that nuclides having paired neutrons tend to have higher relative abundances than those with an odd number of neutrons (Figure 1.4). Thus, the abundant isotopes of elements with even atomic numbers tend to have even mass numbers, but those of elements with odd atomic numbers tend to have odd mass numbers (Appendix A.1). There is at least one stable nuclide for every value of A up to 209, a value that corresponds to the most massive stable nuclide (^{209}Bi). The only exceptions to this are mass numbers 5, 8, 147, and 149, which have none. However, there is never more than one stable nuclide if A is odd, and no more than three if A is even. An example of a mass number sharing three stable nuclides is number 96, composed of the triplet ^{96}Zr, ^{96}Mo, and ^{96}Ru, but in detail the atomic weights of these nuclides are not exactly identical.

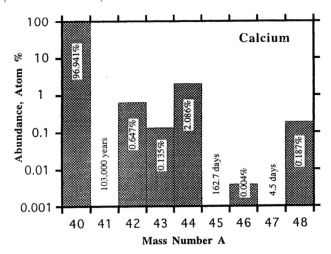

Figure 1.4 Mass spectrum of calcium, showing the preference for isotopes with paired neutrons to be more abundant, if not more stable, than those with odd neutron numbers. An additional preference is seen for isotopes to have mass numbers that are multiples of 4; for example, the isotope ^{40}Ca constitutes 96.94 atom % of all calcium atoms. The only stable, odd-numbered isotope of calcium is ^{43}Ca; the half-lives of the radioisotopes ^{41}Ca, ^{45}Ca, and ^{47}Ca are indicated. All calcium isotopes with $A < 40$ or $A > 48$ are very short-lived, with half-lives of a few minutes or less. Data from Walker et al. (1989).

Thus, not only do the abundant nuclides preferentially contain an even number of protons, they also tend to have an even number of neutrons. More than half of the stable nuclides have this even–even configuration (Table 1.3). In marked contrast, only four stable nuclides, ^{2}H, ^{6}Li, ^{10}B, and ^{14}N, have an odd number of protons along with an odd number of neutrons. While these four nuclides have even mass numbers, those even numbers represent the sum of two odd numbers (equation 1.1). Thus, the even–even preference extends not only to the abundance of the nuclides, but also to their very existence and stability. Such relationships prove that the even–odd effect has a nuclear origin rather than a chemical one (Suess, 1987). These patterns likely relate to the manner in which the elements were formed during the "Big Bang."

Table 1.3 Even–Even Preference for Stable Nuclide Occurrence

A	Z	N	Number of Stable Nuclides	Nuclear Spin
Even	Even	Even	157	0
Odd	Even	Odd	53	1/2, 3/2, 5/2, 7/2
Odd	Odd	Even	50	1/2, 3/2, 5/2, 7/2
Even	Odd	Odd	4	1, 3
Total			264	

After Faure (1986) and Walker et al. (1989).

There is also a stunning preference for nuclides to be abundant and stable if their mass numbers are multiples of 4, the mass of the α-particle. Consider that 85% of the mass of the Earth is composed of only four nuclides, ^{56}Fe, ^{16}O, ^{28}Si, and ^{24}Mg. An additional 3.5 wt.% of Earth is composed of ^{32}S, ^{40}Ca, and ^{60}Ni. A notable exception is mass number 8, which does not exist in nature because it is highly unstable relative to two α-particles.

1.3 Properties and Fractionation of Isotopic Molecules

If the abundances of the stable isotopes were truly invariant, their study would have very limited application. In fact, the abundances given in Table 1.2 are averages, and in detail the values for several elements vary from sample to sample. As a result, the atomic weights of these elements must also vary from sample to sample.

The key question is: What processes cause the isotopes to vary in abundance? Of course, as radioactive nuclides decay over time, radiogenic nuclides are formed, and variations in isotopic abundances are produced in the elements involved. Of more interest here are the processes that cause the stable, nonradiogenic isotopes of certain elements to vary in abundance, or "fractionate." Such processes can be generally characterized as nonequilibrium (kinetic) effects or equilibrium effects.

1.3.1 Nonequilibrium Effects

Isotopic fractionation often accompanies dynamic processes that are fast, incomplete, or unidirectional. Several different processes can give rise to these effects.

1.3.1.1 Diffusion Diffusion can produce abundance variations among the isotopes of any element, be the nuclides stable or not, simply because the various isotopes have different masses. An important example is diffusion of gaseous molecules through an orifice that has an opening smaller than their mean free path. According to the kinetic theory of gases, all molecules at the same temperature have the same average kinetic energy (K.E.). Thus, for two isotopic molecules with molecular weights m_1 and m_2, the respective translational velocities v_1 and v_2 may be computed from the relationship

$$\text{K.E.} = \frac{m_1 v_1^2}{2} = \frac{m_2 v_2^2}{2} \qquad (1.9)$$

This result directly gives rise to "Graham's law of diffusion":

$$\boxed{\frac{v_1}{v_2} = \sqrt{\frac{m_2}{m_1}}} \qquad (1.10)$$

The faster translational velocities of molecules that bear the lighter isotopes will cause their preferential escape through the orifice, leaving a residue that over time becomes progressively enriched in the heavy isotopes.

More sophisticated models may be made with various solutions to the diffusion equation that are appropriate for different situations. In many cases, it is useful to relate the material flux F^* of a trace isotope to the flux F of a common reference isotope, which can commonly be done in terms of the respective diffusion coefficients D^* and D of these isotopes. Commonly, the quotient F^*/F is either proportional to D^*/D, or to the square root of that quantity. Some examples are given in chapter 4.

It is of historical interest that giant, gaseous diffusion plants were employed during World War II to separate fissionable uranium, ^{235}U, from the much more abundant ^{238}U. In order to realize this separation efficiently on an industrial scale, the uranium was first all converted to the gas UF_6, which was then run through diffusion columns several kilometers long.

1.3.1.2 Evaporation   Evaporation is another unidirectional, nonequilibrium process that can cause isotopic fractionation. Though the overall process of evaporation is complex (see chapter 4), one fractionating effect is that the higher translational velocities of molecules containing the "lightest" isotopes may allow them to preferentially break through the liquid surface and escape into the atmosphere. Over time, the residual liquid will typically become progressively enriched in the heavy isotopes.

1.3.1.3 Kinetic Isotope Effects   Kinetic isotope effects generally relate to differences in the dissociation energies of molecules composed of different isotopes. For example, the rate determining step in a set of chemical reactions might involve the breakage of a bond. If it is substantially easier to break the bonds of molecules that contain the lightest isotopes, which is plausible because the vibrational frequency of such bonds will tend to be higher, then the lighter isotopes will be preferentially incorporated in the products of incomplete reactions, while the heavy isotopes will become enriched in the unreacted residue (Bigeleisen, 1965).

1.3.1.4 Metabolic Effects   Organisms commonly also produce nonequilibrium isotopic fractionations. As we breathe, we preferentially utilize ^{16}O for combustion, while ^{17}O and ^{18}O become progressively more abundant in the residual air in our lungs (Epstein and Zeiri, 1988).

Metabolic effects can simultaneously incorporate the effects of several different nonequilibrium fractionations, as well as equilibrium effects. An important example is the process of photosynthesis. As green plants fix carbon, the primary step is the diffusion of CO_2 gas from the atmosphere through the tiny stomata of leaves. Diffusive fractionation occurs at this step, which leads to the preferential incorporation of ^{12}C in the plant. This effect is amplified by subsequent metabolic reactions that can differ among the various plant types. The end result is that all plants have low $^{13}C/^{12}C$ ratios. This characteristic is passed along to herbivores, and ultimately is inherited by other animals in the food chain (e.g., DeNiro and Epstein, 1978), so that the distinctive isotopic signature of photosynthesis is imparted to nearly every familiar life form.

1.3.2 Equilibrium Effects

Isotopic fractionations can also occur in equilibrated systems that comprise two or more phases that contain a common element. In such systems, the atomic ratio R of heavy to light isotopes of the element in common can differ among the equilibrated phases. Equilibrium fractionations tend to be small but can be measured with sensitive mass spectrometers. Quantum mechanical principles can be used to estimate these fractionations in certain cases, most accurately for equilibria among gaseous molecules. In a few cases, these fractionations are large enough to be estimated with classical thermodynamic methods, as discussed in chapter 2.

Equilibrium fractionations are largest for elements that have low atomic weight. They are particularly large for hydrogen, and are significant for carbon, oxygen, nitrogen, silicon, sulfur, and a few other elements. Because the atomic weights of these elements are low, the difference of one or more neutrons translates to a significant relative difference in the masses of the various isotopic forms of the element. These differences are key to the generation of the fractionation effects. This is not the only factor, however. The relative mass difference between the stable nuclides ^{40}Ca and ^{48}Ca is much larger than that between ^{12}C and ^{13}C, yet carbon exhibits significant fractionation effects while calcium generally does not. One reason for this is that fractionation effects are strongly correlated with covalent bonds (O'Neil, 1986a). The atoms in these bonds exhibit vigorous vibrational and rotational motions that are strongly affected by the mass, and the resultant mass-related differences give rise to significant equilibrium fractionation effects. In contrast, atoms that enter into ionic or metallic bonds tend to behave as stationary points, in that they are principally bonded by electrostatic forces that depend strongly on the charge but very little on the mass. Because of this, elements that typically enter into such bonding types tend not to exhibit large fractionation effects.

1.3.2.1 Water   The water molecule exemplifies the phenomenon of equilibrium isotopic fractionation. Because the hydrogen positions in each water molecule may be either D or H, and the oxygen position may be either ^{16}O, ^{17}O, or ^{18}O, water is actually a mixture of nine stable, isotopically distinct molecular species:

$$HH^{16}O \qquad HD^{16}O \qquad DD^{16}O$$
$$HH^{17}O \qquad HD^{17}O \qquad DD^{17}O$$
$$HH^{18}O \qquad HD^{18}O \qquad DD^{18}O$$

A key point is that the properties of these nine molecules are not identical. These differences are not restricted to the obvious differences in mass, but extend to practically all physical and thermodynamic properties. The largest differences are apparent in a comparison of ordinary water and "heavy" water, which dominantly represent $HH^{16}O$ and $DD^{16}O$, respectively (Table 1.4).

Table 1.4 Comparison of the Physical Properties of Ordinary and Heavy Water

Property	H_2O	D_2O
Molar mass[1]	18.01528	20.02748 g
ΔH_v @ 25°C[1]	10.519	10.851 kcal/mol
Vapor pressure @ 25°C[2]	23.756	20.544 torr
Melting point[1]	0.00	3.82°C
Boiling point[1]	100.00	101.42°C
Critical temperature[1]	373.99	370.74°C
Density @ 25°C[1]	0.9970	1.1044
Viscosity @ 25°C[3]	8.93	11.0 millipoise
Disassociation constant[3]	1.0×10^{-14}	1.95×10^{-15}
Latent heat of fusion[3]	1436.3 ± 1 @ 0°C	1515 ± 10 cal/mol @ 3.82°C
Toxicity	None	Poisonous

Sources: [1]Lide, 1991; [2]Pupezin et al., 1972; [3]Kirshenbaum, 1951.

The data in Table 1.4 indicate that the different molecular species of water do not behave identically. As a result, these molecules may be fractionated by a large number of physiochemical processes that depend on any of the properties listed in the table. One important example, discussed in detail in chapters 2 and 3, relates to the higher vapor pressure of H_2O relative to D_2O. This difference leads to the preferential incorporation of D_2O in liquid water relative to that in the water vapor with which it is in equilibrium. This effect, in turn, gives rise to profound hydrogen isotope variations in the atmosphere and hydrosphere (chapter 3).

1.3.3 Isotopic Fractionation Factors

The preceding examples provide a useful means to visualize the general process of isotopic fractionation. "Fractionate" simply means to separate into distinct fractions, or parts. Isotopic fractionation refers to the partitioning of a sample into two (or more) parts that have different ratios of "heavy" (more massive) and "light" (less massive) isotopes than the original ratio. If one of these parts is "enriched" in the heavy isotopes, the other must be "depleted."

A natural way to represent such effects is with the isotopic fractionation factor, referred to by the symbol α. This factor represents the partitioning of isotopes between two separate phases, A and B, or between any two parts of a system, according to the definition

$$\alpha_{A-B} = R_A / R_B \qquad (1.11)$$

where R refers to the atomic ratio N^*/N of a heavy (N^*) to a light (N) isotope. For example, R can represent $^2H/^1H$, $^{13}C/^{12}C$, $^{18}O/^{16}O$, and so on, but while R_A and R_B may have distinct numerical values, in any given equation they necessarily refer to the same two nuclides. Hereafter, the important ratio $^2H/^1H$ is simply denoted D/H.

Isotopic fractionation effects tend to be small, and values for α in most systems are quite close to unity. As an example, ^{18}O and D are both preferentially concentrated in liquid water relative to the water vapor with which it is in equilibrium. For these phases at room temperature, the fractionation factor α for oxygen is 1.0094 while that for hydrogen is 1.079.

In the previous example, the isotopic fractionation effect is larger for hydrogen than for oxygen. Hydrogen generally fractionates more than other elements, mostly reflecting the 2× difference in mass between protium and deuterium. In fact, hydrogen exhibits the largest fractionations observed in nature. One example with a large fractionation is the system in which hydrogen gas is equilibrated with water using a platinum catalyst, where the D/H ratio of the water is 3.7× the ratio in the gas at room temperature. In this case, α is numerically equal to 3.7 and is clearly not close to unity.

For dynamic processes, the quantity α can be used to represent a "separation" factor, defining the difference in isotopic ratio between the bulk system and that of an infinitesimal quantity of material that is being added to or removed from the system at some instant. In certain cases of this type, it is very useful to rewrite equation 1.11 as a differential equation. Thus,

$$d \ln N^* / d \ln N = \alpha \tag{1.12}$$

Depending on the nature of the process involved, α may correspond either to an equilibrium fractionation factor or to a kinetic fractionation factor.

1.4 Material Balance Relationships

It is commonly useful to consider a multicomponent chemical system as the sum of its separate parts, successively labeled here by the subscripts $j = 1, 2, 3$, and so on. The overall isotope ratio of such a system may be derived as follows:

$$R_{system} = \frac{\sum_{j=1}^{n} N_j^*}{\sum_{j=1}^{n} N_j} = \frac{N_1^* + N_2^* + \cdots + N_n^*}{\sum_{j=1}^{n} N_j} \tag{1.13a}$$

This may be written as

$$R_{system} = \frac{N_1}{\sum_{j=1}^{n} N_j} \left(\frac{N_1^*}{N_1} \right) + \frac{N_2}{\sum_{j=1}^{n} N_j} \left(\frac{N_2^*}{N_2} \right) + \cdots \tag{1.13b}$$

This result simplifies to the very important equation

$$R_{system} = \sum_{j=1}^{n} X_j R_j \tag{1.13c}$$

and

$$\sum_{j=1}^{n} X_j = 1 \qquad (1.13\text{d})$$

where X_j refers to mole fractions, calculated in terms of the reference nuclide of interest. This is an exact equation, and it states that the isotope ratio of the entire system may be computed from the sum of the isotope ratios of its components, each multiplied by the appropriate weighting factor X_j. This equation has numerous applications. For example, if two or more different samples are mixed in a bulk ratio defined by X_j, then R_{system} indicates the bulk isotopic ratio of the resultant mixture. In other cases, such as for a redistribution of isotopes among the different phases that constitute a closed isotopic system, the value of R_{system} will be invariant even though the values of R for the individual components may all vary. Also, it will be seen that R is a compact and useful variable for mathematical derivations.

Note that the mole fractions in equations 1.13c and d are specifically defined in terms of the particular reference nuclide of interest. This has several ramifications. First, for fractionated elements, the mole fractions of the reference nuclide need not be identical to the mole fractions of the bulk element. Second, for any given system consisting of distinct parts, the mole fraction of a component defined in terms of one element need not be the same as that expressed in terms of another, nor the same as the fractional mass of that component in the system. These circumstances, while straightforward, are easily overlooked and can lead to considerable confusion.

A great simplification is possible, regarding the first effect, if the minor isotopes are present in very small concentrations. In this case, the reference nuclide constitutes the vast majority of atoms of the element that are present in the system, so that the mole fractions expressed in terms of the reference nuclide are, in fact, essentially identical to those expressed in terms of the bulk element. In natural systems, this substitution will generate negligible error for several elements, such as hydrogen, carbon, nitrogen, oxygen, and so on, because ^{1}H, ^{12}C, ^{14}N, and ^{16}O represent 98.9% or more of the respective atomic abundances (Table 1.2). Similarly, only small errors would obtain for the silicon and sulfur systems, because ^{28}Si and ^{32}S have isotopic abundances of > 92 and > 95 atom %, respectively (Appendix A.1). On the other hand, unacceptable errors could arise for certain other elements, such as chlorine, that contain heavy isotopes in significant abundances.

Regarding the second effect, care needs to be exercised case by case. For the case of a trace isotope in a mixture of n endmembers, the appropriate mole fraction X_k of endmember k may be expressed in terms of the mass fractions M_j of all the various endmembers, and the concentrations C_j of the element of interest in each endmember, as follows:

$$X_k = \frac{C_k M_k}{\sum\limits_{j=1}^{n} C_j M_j}$$

(1.13e)

1.4.1 Mixing Relationship

One of the most important applications of equation 1.13c is its use in the computation of the isotope ratio of simple mixtures. As already mentioned, a complication is that, for any given mixing ratios of two or more components, the values of X_j need not be the same from one isotopic system to another. As a simple example, consider the case where a large freshwater river runs into the ocean. Table 1.5 provides the characteristics of the two hypothetical endmembers, and those of a 50:50 mixture of river water and seawater.

The river water has a very low salinity compared with the seawater. Thus, the salinity of the mixture is the simple average of the endmember values, and, in this case, it is very close to one-half that of the ocean. Similarly, because the oxygen and hydrogen isotopes are directly associated with the water molecule, the appropriate values of X_{river} and X_{sea} are each equal to 0.50, and the isotope ratio of the mixture is simply given by the mean of the endmember values. The calculation is not so obvious for C and Sr. The values for X_{river}^{13} and X_{sea}^{13} are not defined in any simple way by the volumetric mixing ratio of the bulk waters, unless the carbon concentrations of these endmembers are known. Because these mole fractions could, in general, vary anywhere from 0 to 1, the value for the mixture R_{mix}^{13}, could take on *any* value between R_{river}^{13} and R_{sea}^{13} (Figure 1.5). This situation is demonstrated for the strontium system, where assumed numerical values for the endmember concentrations and isotope ratios are explicitly provided for this example. Because the concentration of the Sr is much higher in the seawater than in the river, the Sr concentration in the 50:50 mixture is approximately one-half of that in the ocean, analogous to the situation for the salinity. However, the $^{87}Sr/^{86}Sr$ ratio of the mixture is practically identical to that in the ocean, and is very insensitive to the ratio in the river! As a consequence, on a graph of R versus X for all hypothetical mixtures, the hydrogen and oxygen isotopic systems would give almost perfectly straight lines connecting the endmembers, and so would a plot of R^D versus R^{18} that directly represents one

Table 1.5 Mixing Example

Characteristic	River Water	Seawater	50:50 Mixture
Salinity, g/l	0.22	35.0	17.6
$^{18}O/^{16}O$	0.001995	0.002005	0.002000
D/H	0.000152	0.000156	0.000154
$^{13}C/^{12}C$	R_{river}^{13}	R_{sea}^{13}	R_{river}^{13} to R_{sea}^{13}
$^{87}Sr/^{86}Sr$	0.71200	0.70906	0.70913
Sr, mg/l	0.20	7.70	3.95

Figure 1.5 Graph of the isotope ratio R versus the mass fraction X_2 of component "2" in a simple mixture. The two endmembers plot at opposite corners of the graph and have the ratios R_1 and R_2 for an element of interest. If the concentration of that element is equal in the two endmembers, then binary mixing produces the straight 1:1 line. However, if the elemental concentrations differ between the endmembers, then the isotope ratio of the mixtures is a function of the concentration ratio, C_2/C_1, that is indicated on the mixing curves.

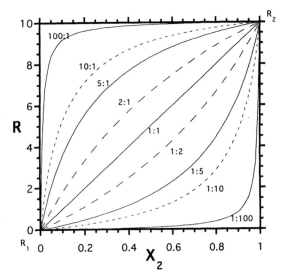

isotopic system relative to another. In contrast, the carbon and strontium systems would yield hyperbolas whose curvature would, in each case, depend on the end-member ionic concentrations of the DIC (dissolved inorganic carbon) and the Sr (Figure 1.5).

In such systems, it is commonly desirable to directly graph the isotope ratio of one element versus that of another. For a simple, nonfractionating, mixing process between two endmembers denoted by "A" and "B," the various isotopic ratios of the resultant mixture ("Mix") are defined by the equation (R. E. C., unpublished)

$$\left[\frac{C_B}{C_A}\left(\frac{R^\dagger_{\text{Mix}} - R^\dagger_B}{R^\dagger_{\text{Mix}} - R^\dagger_A}\right)\right]_{\text{Element } L} = \left[\frac{C_B}{C_A}\left(\frac{R^\dagger_{\text{Mix}} - R^\dagger_B}{R^\dagger_{\text{Mix}} - R^\dagger_A}\right)\right]_{\text{Element } M}$$

$$= \left[\frac{C_B}{C_A}\left(\frac{R^\dagger_{\text{Mix}} - R^\dagger_B}{R^\dagger_{\text{Mix}} - R^\dagger_A}\right)\right]_{\text{Element } N} = \cdots$$

(1.14a)

where C_A and C_B represent the concentrations of the bulk element of interest in the endmembers denoted by the subscripts. Note that each term is, with a change in sign, equal to the bulk mixing ratio defined by the masses of the two end-members A and B that are mixed; that is,

$$\left[\left(\frac{C_B}{C_A}\right)\left(\frac{R^\dagger_{\text{Mix}} - R^\dagger_B}{R^\dagger_{\text{Mix}} - R^\dagger_A}\right)\right]_{\text{Element } L} = \left[\left(\frac{C_B}{C_A}\right)\left(\frac{R^\dagger_{\text{Mix}} - R^\dagger_B}{R^\dagger_{\text{Mix}} - R^\dagger_A}\right)\right]_{\text{Element } M}$$

$$= \cdots = -\frac{\text{Mass } A}{\text{Mass } B}$$

(1.14b)

Also, for equation 1.14a, b to be exact, note that the definition of $R^\dagger$ is different in detail than that of the standard isotope ratio R. Specifically, in equations 1.14a and b,

$$R^\dagger = \frac{\text{mass of isotope}}{\text{total mass of element}} \qquad (1.14c)$$

Because D, ^{13}C, and ^{18}O are trace isotopes, negligible error will result for the H, C, and O isotopic systems if the usual isotope ratios are simply substituted for $R^\dagger$ in equation 1.14. However, the error will be much worse if certain other isotope ratios—for example, ^{87}Sr/^{86}Sr—are substituted for $R^\dagger$. Completely unacceptable errors will result if certain other isotopic ratios—for example ^{206}Pb/^{204}Pb or ^{37}Cl/^{35}Cl—are substituted for $R^\dagger$. For such elements, equation 1.14a may still be used, but the appropriate values for $R^\dagger$—for example, ^{206}Pb/[total Pb]— must be used as defined in equation 1.14c. After these values for $R^\dagger$ are explicitly calculated from equation 1.14c, they may be used in a subsequent calculation to determine the actual isotope ratios of interest.

1.5 Mass Spectrometers

1.5.1 Elementary Physics

For numerous situations, mass spectrometry is the method of choice for determining precise chemical or isotopic abundances. This technique resolves a given sample into its "mass spectrum," where the latter represents the group of components distinguished in terms of mass. The prototype of this device, constructed by Thomson and improved by Aston, was used to discover the isotopes of neon, and subsequently employed to discover numerous other nuclides. These early instruments, actually called "mass spectrographs," utilized either phosphorescent screens or photographic plates to detect the different ion beams. While this technology was sufficient to detect the existence of many substances, quantitative determination of their abundances was very difficult. Thus, the mass spectrograph evolved into the modern "mass spectrometer," which features quantitative electronic detection of ions.

In order to achieve this, the substance or element of interest is first converted into a form suitable for ionization inside a high-vacuum chamber. Following ionization, the substance is accelerated through a high voltage difference, and then collimating slits and electronic lenses are employed to focus the ions into a narrow beam. The charged, energetic beam is passed through strong electric and/ or magnetic fields, which serve as a "prism" in that they resolve the single collimated beam into several distinct beams. In particular, each beam is composed of ions that have a common characteristic, such as momentum, that differs from that character of the ions in another beam. The individual beams that constitute the "mass spectrum" may then be electronically counted and compared, ultimately to be translated into useful terms, such as relative abundances.

An elementary physical analysis of the mass spectrometer can be made in terms of an energy equation and a force (momentum) equation. The kinetic energy of the particles in the ion beam, which are nonrelativistic in most spectrometers, is equal to the charge q multiplied by the potential drop V across the high-voltage source, giving the classical result,

$$qV = mv^2/2 \qquad (1.15)$$

where m is the particle mass and v is the particle velocity. The moving, charged particles are then affected by magnetic (B) and/or electric (E) forces in the spectrometer, according to Lorentz's law:

$$|\vec{F}| = q\vec{E} + q\vec{v} \times \vec{B} \qquad (1.16)$$

According to the configuration of the instrument, the following focusing conditions can be achieved.

1.5.1.1 Velocity Focusing   In this configuration, the magnitudes of crossed E and B fields may be adjusted so that there is no net deflection of the collimated ion beam from its straight path (Figure 1.6a). In this case, there is no net force on the ions, so that $F = 0$ and equation 1.16 becomes

$$v = E/B \qquad (1.17a)$$

This condition is known as "velocity focusing" because only particles with the velocity E/B will emerge through the system undeflected. Thus, velocity focusing may be used along the ion path to remove undesired constituents from the charged beam.

1.5.1.2 Energy Focusing   In some instruments, an analyzer is constructed of curved electrostatic plates (Figure 1.6b). The radius of curvature of the ions may be defined by equating the electrostatic force, qE, with the centripetal force, mv^2/r, that corresponds to the circular motion of the ions. As a result,

$$r = \frac{mv^2}{qE} \qquad (1.17b)$$

Because the radius is directly proportional to the classical kinetic energy of the ions, $mv^2/2$, this condition is known as "energy focusing." In particular, only particles that have a kinetic energy equal to $qEr/2$ can move along the path of radius r. These selected ions are therefore able to transit through the slot between the curved plates, while particles that have different energies will strike the walls. Many instruments, including most ion probes, employ this type of electrostatic analyzer to refine the particle beam by removing undesired components.

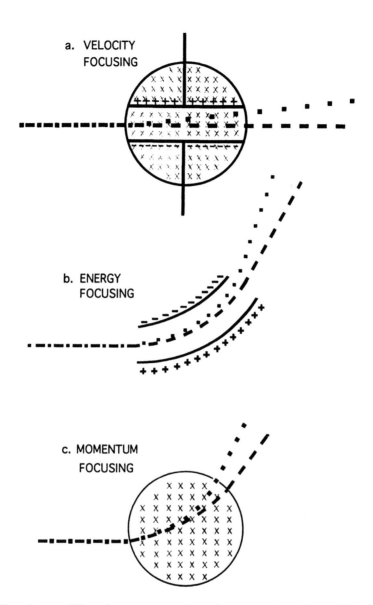

Figure 1.6 Focusing conditions for mass separation of a composite, collimated ion beam consisting of a heavy (dash) and a light (dot) isotope. Electric fields are denoted by the charged metal plates (heavy lines with + and − charges) and are oriented parallel to the page, while the magnetic fields (circled ×s) are oriented into the page. (a) Velocity focusing: crossed electric and magnetic fields allow only ions with the velocity E/B to pass through the plates undeflected (equation 1.17a). (b) Energy focusing: curved, charged electric plates separate ions according to their energy (equation 1.17b). (c) Momentum focusing: a charged ion moves along a circular path in a magnetic field. Ions that have the greatest momentum move along the path that has the least curvature (equation 1.17c).

1.5.1.3 Momentum Focusing   Numerous instruments, including those most commonly used to measure stable isotopes, employ a simple magnetic analyzer to separate the ion beam into its isotopic components (Figure 1.6c). Charged particles moving through a magnetic field encounter a force of magnitude $qv \times B$ (equation 1.16) that imparts a centripetal force mv^2/r of equal magnitude to the particles. For an orthogonally oriented B field, the ions follow a circular path of radius r, where

$$r = mv/(qB) \qquad (1.17c)$$

This creates the condition known as "momentum focusing." In particular, particles with the same charge q and same momentum mv will all move with the same radius of curvature, while those that have different values will define different paths. It is useful to use equation 1.15 to express this latter result in terms of the accelerating voltage V and magnetic field strength B of the instrument, and the charge-to-mass ratio q/m of the particular ion, as follows:

$$r = \sqrt{\frac{2Vm}{B^2q}} \qquad (1.17d)$$

According to equation 1.17d, ions that have different values of q/m will move along paths that have different radii, and therefore will separate from each other. That is, for a given charge q, the more massive the ion, the larger the radius of curvature. This effect gives rise to the different ion beams in the instrument, due to their different masses. However, a doubly ionized ion will focus at the same place as a singly ionized particle that has half the mass; for example, CO_2^{2+} (mass 44) would focus at essentially the same place as $^{22}Ne^+$.

1.5.2 Isotope Ratio Mass Spectrometry

Mass spectrometers configured to make precise isotopic determinations invariably have three basic constituents, here listed sequentially along the travel path of the particles. First, an efficient *source* is used to generate, accelerate, and collimate the ions. In the *analyzer*, strong electric or magnetic fields are used to separate the ion beam into its different constituents, according to one or more of the focusing conditions discussed previously. Last is the *collector*, where the different ion beams are electronically counted or compared. All of these components may be optimized to achieve different analytical purposes.

The most common type of mass spectrometers used to measure ratios of light stable isotopes are configured to analyze gaseous samples. In most cases, these instruments employ a magnetic sector alone (equations 1.17c and d) for the resolution of the ion beam into its components (Figure 1.7). The basic design of this type of instrument was pioneered by A. O. Nier (1947) and used to determine the precise isotopic abundances of many materials, including Ar and N_2 in air, as well as CO_2 and H_2.

The gas mass spectrometer contains many features that permit precise determinations of the small abundances and the small variations in the light stable

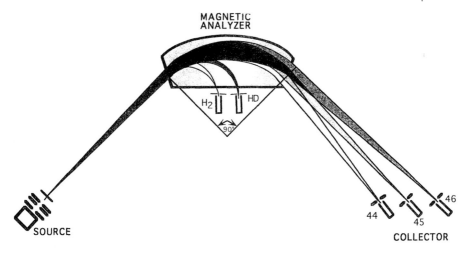

Figure I.7 Diagram of the MAT 252 mass spectrometer, indicating the source, analyzer, and collector. The source generates and collimates singly charged (+1) ions that have a uniform energy of 10 kV, while the analyzer is a magnetic field that separates the ions according to their momenta. The trajectories of CO_2 with masses 44, 45, and 46, and of H_2 and HD, are shown.

isotopes. First, a dual inlet system allows for introduction of the gas into the high-vacuum source chamber that in most cases is continuously pumped. The gas passes through long (>1 m) capillaries that have a diameter (typically 10^{-4} m or 0.1 mm) that is 1000× greater than the mean free path of the gas molecules (ca. 10^{-7} m). Such capillaries permit the inlet pressures to be relatively high, creating a steady mass flow of gas sufficient to prevent back diffusion, and to achieve the nonfractionating condition permitted by viscous flow. A changeover valve allows for the quick interruption of flow of the sample gas and its replacement with a standard gas, thus permitting rapid and repeated comparison of the sample with the standard.

Once in the source, an "electron gun," typically composed of a hot (2000°C) tungsten filament, is used to produce a small (~1 mA) current of electrons with an energy of about 90 eV. These electrons are sufficiently energetic to ionize the gas molecules by bombardment, in that the first ionization energies of the molecules are lower. For example, the first ionization potential of CO_2 is 13.77 V and that of H_2 is 15.43 V. The source conditions are optimized to singly ionize a "significant" fraction of the particles, which in reality may be only one in several thousand of them. Next, charged plates in the source chamber accelerate the ionized gas to nonrelativistic energies of about 10 kV; then, collimating plates and electronic lenses are used to focus the ions into a narrow (<1 mm) beam.

In the analyzer, the ion beam encounters a magnetic field, produced by either a permanent magnet or an electromagnet, that can be as strong as 1 tesla (1 T = 10,000 gauss) or more. The required strength of the magnet generally increases as the molecular weight of the gas being analyzed is increased (equation 1.17d). The magnetic field causes the ions to move along circular paths that depend on their momenta (equation 1.17c). By the time the beam leaves the magnetic sector, it has been deflected by a significant angle that is commonly 60°, 90°, or even 180° in some instruments; and, in detail, the beam has been resolved into its different isotopic constituents. Numerous design details of the magnetic fields are used to increase the separation of the resolved beams.

Placed at the terminus of the ion path, the collector quantifies the ion beams by detecting and measuring the tiny currents generated by their neutralization. This may involve one or, more typically, several different detectors, such as "Faraday cups," electron multipliers, and so on. If only one detector is used then beam scanning is required, while multiple detectors permit the simultaneous measurement of two or more isotope ratios from a stationary beam condition. Many modern detectors utilize solid-state devices, such as "voltage-frequency converters," that change voltages to digital readouts in hertz, resulting in pulses that are fed to counters for a selected "integration time." Precise measurements of isotopic abundances can now be routinely made on a few micromoles of gas, and even on much smaller samples with special effort.

It is useful to point out that when the first spectrometers were designed for stable isotope abundance determinations, such solid-state devices were not available and direct counting of the small beams was not possible. This difficulty was originally overcome by Nier (1947) through the use of a "potential divider," analogous to the principle of the "Wheatstone Bridge." The use of voltages is key here, because the tiny ion currents, when passed through very large resistors, result in appreciable and easily measured voltage drops, according to Ohm's law. In practice, the intensity of the major isotope ion current of the sample was adjusted to be the same as that of the reference standard. This was achieved by alternately introducing the sample and the standard gases into the spectrometer, passing the respective ion currents through a large resistor, and adjusting the pressure on the sample gas until the sample voltage drop became identical to that of the standard. Next, the tiny ion current that represented the trace isotope beam was run through a very large resistor (e.g., 10^{11} Ω), perhaps 100–1000× greater than that used for the major ion beams that represented the abundant isotope. Smaller resistors were then added until the voltages from the major and trace ion beams, representing the product of the ion currents multiplied by the appropriate resistances, became identical. In the simplest case, if these voltages are balanced exactly both for the sample and then for the standard, the isotope ratios can be defined in terms of the resistances needed to balance these voltages for the sample (Ω_x) and the standard (Ω_{std}) gases:

$$R_x/R_{std} = \Omega_{std}/\Omega_x$$

(1.18a)

This straightforward technology, though cumbersome and slow by today's standards, permitted many excellent determinations of isotopic abundances to be made more than 60 years ago!

1.5.3 Ion Probes and Other Isotope Ratio Mass Spectrometers

Mass spectrometry is a highly developed field, and numerous different instruments have been devised for particular problems. Applications range from rapid chemical characterization of bulk samples, to analyses of microscopic spots in semiconductors, to precise isotopic abundance determinations of trace elements. One spectacularly successful mass spectrometer, small enough to fit into a briefcase, was launched to determine the composition of the Martian atmosphere (Nier, 1989).

The most important differences in mass spectrometers may be seen in terms of the design of their sources or of their analyzers. The source is generally one of two types, depending on whether the instrument is a gas mass spectrometer or a thermal ionization mass spectrometer. The first type is used for most measurements of light stable isotopes, as well as for noble gas studies, and depending on the sample size may be dynamically pumped during analysis or statically analyzed after it is introduced into the spectrometer. The thermal ionization method is used for many metals, such as Sr, U, and so on, that are coated on, and ultimately vaporized from, a heated filament in the source chamber.

In addition to these spectrometers, several new types have been devised to exploit other means of introducing samples to the analyzer. These include secondary ion mass spectrometers (ion microprobes), discussed below; resonance ion mass spectrometers that feature use of a narrow-band laser to selectively ionize desired constituents from a sample; and ICP-MS (inductively coupled plasma mass spectrometers) that ionize bulk samples in a plasma, simultaneously rendering practically all the elements in a form suitable for analysis. Yet another type is accelerator mass spectrometry, which utilizes extremely high energies (e.g., 10 MeV) generated in particle accelerators to achieve extremely high sensitivity, permitting the analysis of certain nuclides (e.g., T, ^{14}C, ^{36}Cl, etc.) that are so rare that they constitute only a few thousand atoms in an entire sample. This technique has fostered the development of an entirely new scientific field that mostly exploits the new ability to precisely measure short-lived cosmogenic radionuclides that are continually produced by cosmic rays hitting the atmosphere and terrestrial and extraterrestrial materials.

Of all these new techniques, the ion probe has made the greatest contribution to stable isotope studies (Figure 1.8). In this technique, secondary ions are produced from a sample by bombardment with heavy ions, such as cesium or oxygen. Because of the nonselectivity of the types of nuclides that are introduced into the spectrometer—which results in a need to eliminate isobaric interferences, such as element hydrides—this type of instrument requires extremely high resolution. This is achieved by "double focusing," in that ion probes employ both electrostatic and magnetic analyzers. The mass resolving power (MRP) is defined as

$$\boxed{\text{MRP} = M/\Delta M} \qquad (1.18b)$$

where M is the mass of the nuclide of interest and ΔM refers to the peak separation. For ion probes, the MRP may be as high as 40,000, a value more than 100×

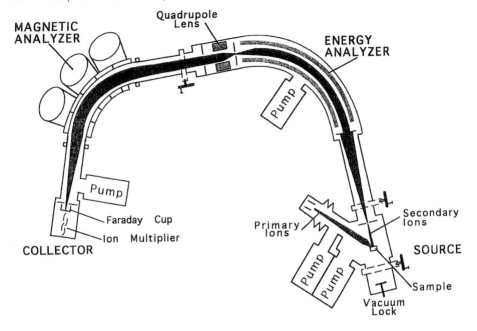

Figure 1.8 Diagram of the SHRIMP ion microprobe, a versatile instrument with very high mass resolution. In the primary source, energetic ions such as Cs^+ are generated that bombard the sample, sputtering off a secondary beam that is accelerated and collimated. The beam then passes through an energy analyzer, consisting of curved electrostatic plates, then through a magnetic analyzer that selects for momentum, and ultimately to the collector for the quantitative detection of the ions. After Eldridge et al. (1989).

greater that that used in many stable isotope mass spectrometers. Moreover, these instruments attain this high resolution without large transmission losses, and therefore have very high sensitivity, enabling measurements of concentrations less than 1 ppb for many constituents. The ion probe can also measure extremely small samples (to 10^{-15} g), and in many cases can be set up to generate spot analyses on the surface of larger samples. The ion probe is very good for trace element measurements and for the examination of small-scale isotopic variations of certain elements. Its most successful applications to stable isotopes have been for the sulfur system (e.g., Eldridge et al., 1989) and in the elucidation of isotopic anomalies in extraterrestrial materials (e.g., Zinner, 1989).

1.6 Notation and Standards

1.6.1 Delta Notation

The isotope ratio R is a very useful variable for reporting data for certain isotopic systems, particularly for those that involve radiogenic nuclides (e.g., $^{87}Sr/^{86}Sr$, $^{206}Pb/^{204}Pb$, etc.). However, R is generally not used to report stable isotope measurements because the absolute abundances of the heavy isotopes are, in many cases, very low (e.g., $D/H \sim 0.00015$). Moreover, the natural variations in the abundance of the stable nonradiogenic isotopes are normally quite small. As a result, routine measurements of H, C, and O isotopes are made by electronically counting and comparing the intensities of the resolved beams, then employing a switching mechanism that allows rapid comparison of the intensities of the sample beams with those of a known gas standard. This method permits the isotopic constitution of the unknown to be determined by the *difference* in intensities from those of the standard, and circumvents the necessity of making measurements of absolute intensities for each sample, which would be more difficult.

For all the above reasons, for many stable isotope systems it is natural to directly report the measured difference in the isotopic composition of the sample (*x*) and an accepted standard (std) in terms of dimensionless δ-values (termed "delta-values"), defined by the formula

$$\delta_x = 1000 \frac{R_x - R_{std}}{R_{std}}$$

(1.19)

where the R values refer to the isotope ratios, either D/H, $^{13}C/^{12}C$, or $^{18}O/^{16}O$, as appropriate. Depending on the element of interest, this formula defines the δD (or δ^2H), $\delta^{13}C$, or $\delta^{18}O$ values. The factor of 1000 converts the δ-values to per mil (‰). Less commonly, a factor of 100 is used to report hydrogen isotope data, in which case the δD values are given in per cent (%). For these different cases, equation 1.19 specifically becomes one of the following:

$$\delta D = 1000 \left[\frac{\left(\frac{^2H}{^1H}\right)_x}{\left(\frac{^2H}{^1H}\right)_{SMOW}} - 1 \right]$$

(1.20a)

$$\delta^{18}O = 1000 \left[\frac{\left(\frac{^{18}O}{^{16}O}\right)_x}{\left(\frac{^{18}O}{^{16}O}\right)_{SMOW}} - 1 \right]$$

(1.20b)

$$\delta^{13}C = 1000 \left[\frac{\left(\frac{^{13}C}{^{13}C}\right)_x}{\left(\frac{^{13}C}{^{13}C}\right)_{PDB}} - 1 \right]$$

(1.20c)

The isotopic standards used in equations 1.20a, b, and c are those most widely used, and are defined in the following section.

1.6.2 Isotopic Standards

Isotopic standards are chosen for convenience, and, of course, are assigned the value of zero per mil on the δ-scale of interest (equation 1.19). Ideally, an isotopic standard would be a homogeneous material, with an isotopic ratio similar to those of the samples undergoing investigation, that is available in large quantity, easy to distribute, and safe and easy to process. Historically, this ideal has been rarely attained, and many important materials are either inhomogeneous or are no longer available, and so forth. Nevertheless, useful isotopic reference materials are distributed by the National Bureau of Standards, and others are available commercially.

The most useful standard for oxygen and hydrogen isotopes is standard mean ocean water (SMOW), defined by Craig (1961). The most common standard for carbon and for oxygen isotopes in carbonates is referred to as "PDB," and represents a Cretaceous belemnite from the Peedee Formation of South Carolina (Craig, 1957). The $^{15}N/^{14}N$ ratio of air is generally used to calibrate nitrogen isotope data, while the $^{34}S/^{32}S$ ratio of troilite from the Canyon Diablo meteorite (Meteor Crater, Arizona) is used for sulfur studies. Several secondary standards have been calibrated relative to these primary standards, and, of course, other materials are used for other isotopic systems. More detailed discussions are provided by Friedman and O'Neil (1977), O'Neil (1986b), and Hoefs (1987).

Attempts are sometimes made to determine the absolute isotopic abundances of these standards. For example, Craig (1961) determined the absolute abundances of ^{18}O in SMOW to be 1989.5 ± 2.5 ppm and that of D to be 158 ± 2 ppm. Subsequent measurements generally, but do not exactly, corroborate Craig's results. Thus, three more recent studies found D concentrations of 155.7 ± 0.1 ppm in SMOW, and another study reported 2005.2 ± 0.45 for ^{18}O (see review by O'Neil, 1986b). If the analytical differences between these determinations were expressed in per mil, they would be nearly 15‰ for the D and 8‰ for the ^{18}O in SMOW, respectively! These uncertainties are *much* larger than the precisions, usually better than ±1‰ for D and ±0.15‰ for ^{18}O, that can be routinely attained in a modern laboratory that reports analyses in terms of their difference from a defined, yet not precisely known, isotopic standard. This example illustrates that it is much more difficult to determine the absolute concentration of a trace isotope in a material than it is to define its isotopic difference from another material.

1.6.3 Standard Conversion Identity

It is commonly necessary to convert $\delta_1 X$, the δ-value of a sample ("X") measured relative to one isotopic standard ("1"), to the value $\delta_2 X$ which would be measured relative to a second standard ("2"). This conversion is approximately made by adding the per mil difference, $\delta_2 1$, between the two standards, but is exactly accomplished with the standard conversion identity (Craig, 1957):

$$\delta_2 X = \delta_1 X + \delta_2 1 + \tfrac{1}{1000}(\delta_1 X)(\delta_2 1) \qquad (1.21)$$

For example, the $^{18}O/^{16}O$ ratio of the Peedee belemnite is commonly used as an alternative $^{18}O/^{16}O$ standard for oxygen extracted from carbonates. This oxygen standard is also called PDB, but will not be utilized in this book. A $\delta^{18}O$ measurement of calcite reported on the PDB scale may be readily converted to the SMOW scale, or vice versa, with the relationship derived from equation 1.21 (Friedman and O'Neil, 1977; O'Neil, 1986b):

$$(\delta^{18}O_x)_{SMOW} = 1.03086(\delta^{18}O_x)_{PDB} + 30.86 \qquad (1.22)$$

Note that the standard conversion identity (equation 1.21) is equivalent to the statement

$$\boxed{\beta = \frac{1000 + \delta_2 X}{1000 + \delta_1 X}} \qquad (1.23a)$$

where the conversion factor β is equal to the quotient of the absolute isotope ratios of the two standard materials, R_1/R_2. Alternatively,

$$\beta = \frac{1000 + \delta_2 1}{1000} \qquad (1.23b)$$

1.6.4 Approximations

The quantity $\delta_A - \delta_B$ is defined as Δ_{A-B} and is called "Big Delta." This quantity provides a convenient but rough estimate of the "fractionation" between the two phases A and B:

$$\Delta_{A-B} = \delta_A - \delta_B \qquad (1.24a)$$

For example, if δ_A were $+15.0$ and δ_B were -5.0 per mil relative to some specified standard, a geochemist would state that "there is a 20 per mil fractionation between phases A and B," but the actual value of α_{A-B} would be close to 1.0201. This convenient situation arises because isotopic fractionation factors are normally very close to unity. In such cases, it is easy to demonstrate that the following approximations are reasonably accurate:

$$\Delta_{A-B} \cong 1000(\alpha_{A-B} - 1) \qquad (1.24b)$$

and

$$\Delta_{A-B} \cong 1000 \ln \alpha_{A-B} \qquad (1.24c)$$

Because of the latter relationships, some authors define the quantity Δ_{A-B} as being equal either to the quantity $1000(\alpha_{A-B} - 1)$ or to the quantity $1000 \ln \alpha_{A-B}$, but the definition given in equation 1.24a is preferred here. Other authors occasionally use the definition $\varepsilon = 1000(\alpha - 1)$; this is not to be confused with the "epsilon units" used in Nd isotope studies, and in any case will not be employed in this book.

1.6.5 Fundamental Equations

Because of their great importance, it is useful to convert the fundamental equations of isotopic fractionation and of material balance in terms of δ-values. This can be readily accomplished by using equation 1.19 to modify equation 1.11 to implicitly express the relationship between α and δ; specifically,

$$\alpha_{A-B} = \frac{1000 + \delta_A}{1000 + \delta_B} \qquad (1.25)$$

where δ_A and δ_B refer to the δ-values for two samples A and B, provided, of course, that they have been measured relative to the same isotopic standard. Equation 1.25 is an important and mathematically exact relation, which can be used to calculate the apparent (i.e., measured) fractionation factor from the δ-values of two associated phases, or to predict the δ-value of one phase from that of another if equilibrium has been attained and if the appropriate fractionation factor is known.

Conservation of mass in an n-component isotopic system may also be expressed in terms of δ-values. Equation 1.13 becomes

$$\delta_{\text{system}} = \sum_{j=1}^{n} X_j \delta_j \qquad (1.26a)$$

where, again,

$$\sum_{j=1}^{n} X_j = 1 \qquad (1.26b)$$

where X_j and δ_j, respectively, represent the mole fractions and δ-values for each constituent phase. The X_j values are exactly the same quantities that were discussed previously.

1.7 Summary

Atoms are composed of protons, neutrons, and electrons. The number of neutrons can vary among the various atoms that constitute a given element, giving rise to the phenomenon of "isotopes." Literally, isotopes are atoms of a single element that have different atomic weights. Approximately half of the elements have two or more stable nuclides, although a few common elements, including F, Na, and Al, have only one.

The distribution and abundance patterns of the nuclides are of great interest. Even-numbered elements, and particularly isotopes with even mass numbers, tend to be much more abundant in nature than those that are odd-numbered. Certain stable atoms, called "radiogenic" isotopes, are produced by the decay of radioactive parent isotopes, and these radiogenic isotopes exhibit profound abundance variations that can be used to determine the ages of geologic materials. In contrast, abundance variations of the nonradiogenic stable isotopes tend to be small, and are produced by diverse processes, collectively called "isotopic fractionations," that arise from the mass differences among the various isotopes. These fractionations may accompany kinetic processes such as diffusion, equilibrium processes, and metabolic processes. The magnitude of these processes is indicated by the "isotopic fractionation factor," α, which expresses the quotient of the respective isotopic ratios of two different materials.

Because the variations in the abundance of the stable nonradiogenic isotopes are normally rather small, sensitive devices (called isotope ratio mass spectrometers) have been devised to make precise determinations of their abundances. These instruments generally ionize, accelerate, and collimate the particles and pass them through magnetic or electric fields that cause the separation of the different masses. Different configurations of the magnetic and electric fields in mass spectrometers may be used to effect the separation of different isotopes by distinguishing the velocities, momenta, or energies of the particles. Mass spectrometers may be optimized to facilitate many different types of measurements.

For many reasons, abundances of the light, stable, nonradiogenic isotopes are generally reported as "delta-values." These values are not absolute measurements of abundances, but rather are precise determinations of the difference in the isotope ratios between the sample and a specified standard material.

Two key equations, one expressing the isotope fractionation factor and the other representing material balance of the isotopes, may be used to solve diverse problems in the physical sciences. These basic laws may be readily translated from their representation in terms of isotope ratios to the δ-notation.

1.8 Problems

1. How many protons, neutrons, and electrons are present in neutral atoms of ^{37}Cl and ^{208}Pb, and in the cation $^{88}Sr^{2+}$?

2. Use equation 1.2 to calculate the ionization energy of an unexcited hydrogen atom. You will need to employ the relation $E = hc/\lambda$, where E is the energy, h is Planck's constant, and c is the speed of light (See Appendix A.2.). Compare your result with the observed value of 13.58 eV.

3. Use equation 1.3 to calculate the wavelength of Cu $K\alpha$ radiation, which is widely used in X-ray diffractometers. Compare your result with the result given by equation 1.4, and also with the observed wavelength of 1.54 Å.

4. Calculate the atomic weight of hydrogen in the water used to cool a nuclear reactor, if the coolant is (a) pure H_2O, (b) pure HDO, and (c) pure D_2O.

5. Use the approximate isotopic abundances for carbon and oxygen given in Table 1.2 to calculate the approximate relative abundances of CO_2 masses 44, 45, and 46 in a mass spectrometer.

6. (a) Calculate the atomic weight of uranium, given that the mass of ^{238}U is 238.050785 amu, the mass of ^{235}U is 235.043924 amu, and the present-day atomic ratio of ^{238}U to ^{235}U is 137.88.

(b) Calculate the atomic weight of uranium when Earth formed 4.55 Ga ago, given the above information, and given that the half-life of ^{238}U is 4.468 Ga, and the half-life of ^{235}U is 0.7038 Ga.

7. Suggest several reasons why it is advantageous to use UF_6 to effect the separation of uranium isotopes.

8. A river with a $\delta^{18}O$ value of $-7.0‰$ has a salinity of 0.1 g/l and contains 200 mg/l of dissolved bicarbonate with a $\delta^{13}C$ value of $-10.0‰$. This water enters and mixes with surface waters of the ocean, which have a $\delta^{18}O$ value of $+0.5$, a salinity of 35.0 g/l, and contain 120 mg/l of dissolved bicarbonate with a $\delta^{13}C$ value of $+1.0$. If 1 kg of river water is mixed with 2 kg of seawater, what are the $\delta^{18}O$ and the $\delta^{13}C$ values of the mixture? Plot a graph of the $\delta^{18}O$ and the $\delta^{13}C$ values versus salinity in the mixing zone.

9. From 1956 to 1988, the atmospheric CO_2 concentration increased from about 314 to 351 ppmv, while the $\delta^{13}C$ value decreased from $-6.7‰$ to $-7.8‰$ (Keeling et al., 1989). What would the isotopic shift have been, had the increase in concentration been caused by the simple addition of carbon dioxide produced by the combustion of fossil fuels, which has a value close to $-27.0‰$? Suggest a reason for any discrepancy.

10. Calculate the $\delta^{18}O$ values of the following samples relative to SMOW: (a) pure $H_2{}^{18}O$; (b) pure $H_2{}^{16}O$; (c) a 50:50 mix of (a) and (b); and (d) a water sample with an isotope ratio twice as large as that of the SMOW standard.

11. Use the definitions of α and δ to prove that the isotopic fractionation factor between two phases A and B is equal to

$$\alpha_{A-B} = \frac{1000 + \delta_A}{1000 + \delta_B}$$

provided that δ_A and δ_B have been measured relative to the same isotopic standard.

12. (a) Prove the following identity:

$$\alpha_{A-B} - 1 = \left(\frac{\delta_A - \delta_B}{1000 + \delta_B}\right)$$

(b) Use your result to discuss the circumstances when the following approximation is an accurate one:

$$\delta_A - \delta_B \cong 1000(\alpha_{A-B} - 1)$$

13. Use a MacLaurin series or similar argument to prove that, for values of α near unity,

$$\ln \alpha \cong (\alpha - 1)$$

14. Use equation 1.17d to compute the radius of curvature for a singly ionized beam of CO_2 in the analyzer of a MAT 251 mass spectrometer that has a magnetic field of 0.37 T, and a source accelerating voltage of 10 keV. *Answer*: 0.26 m.

15. Calculate the minimum mass resolving power needed for a gas mass spectrometer to distinguish the various isotopic species of CO_2. Compare this result with the MRP required to separate ^{48}Ti (47.9480 u) and ^{48}Ca (47.9525 u), which may be accomplished with ion probes. What MRP would be needed to separate ^{87}Rb (86.909188 u) and ^{87}Sr (86.908884 u), a feat that is currently unattainable?

16. What "β-factor" is needed in equation 1.23a to compare oxygen isotopic measurements from the SMOW and the PDB scales? If the $\delta^{18}O$ value of a shell is $-10.3‰$ on the PDB scale, what is its value relative to the SMOW standard?

17. Calculate the velocity and kinetic energy (E_k) of a singly ionized, mass 44, CO_2^+ ion that has been accelerated across the 10 kV voltage drop in the source of a mass spectrometer. Compare this energy with the rest energy ($E_0 = mc^2$) of the ion. Given the rule of thumb that relativistic corrections are needed if the ratio E_k/E_0 is more that 1:100, does the instrument operate in the "classical" region?

18. How many mass 44 ions are being counted per second by a mass spectrometer under normal operating conditions, if that ion beam gives a voltage of 3.0 V across a resistor of $3 \times 10^8 \, \Omega$? If the ionization efficiency of the source is 1:2300, how many ions per second are being transited through the 0.1-mm diameter capillary tubes, and what is their velocity ?

19. Calculate the mean free path of carbon dioxide at 298 K, and compare this result with the 0.1-mm diameter capillary tube. Does the instrument operate in the molecular flow or viscous flow region?

20. Calculate the "raw" $\delta(45)$ and $\delta(46)$ values of the CO_2 gas sample from the voltages provided, which have already been corrected for background "noise." What correction is necessary to convert those values to $\delta^{13}C$ and $\delta^{18}O$ values relative to the machine standard? What additional information is then needed to correct the values to the PDB and SMOW standards?

	Mass 44	Mass 45	Mass 46
Sample voltages	1.483	1.750	2.049
Standard voltages	1.517	1.818	2.125
Resistor	$3 \times 10^8 \, \Omega$	$3 \times 10^{10} \, \Omega$	$1 \times 10^{11} \, \Omega$

References

Bernatowicz, T., Brannon, J., Brazzle, R., Cowsik, R., Hohenberg, C., and Podosek, F. (1992) Neutrino mass limits from a precise determination of $\beta\beta$-decay rates of ^{128}Te and ^{130}Te. *Phys. Rev. Lett.*, **69**, 2341–2344.

Bigeleisen, J. (1965) Chemistry of isotopes. *Science*, **147**, 463–471.

Craig, H. (1957) Isotopic standards for carbon and oxygen and correction factors for mass-spectrometric analysis of carbon dioxide. *Geochim. Cosmochim. Acta*, **12**, 133–149.

Craig, H. (1961) Standard for reporting concentrations of deuterium and oxygen-18 in natural waters. *Science*, **133**, 1833–1834.

DeNiro, M.J. and Epstein, S. (1978) Influence of diet on the distribution of carbon isotopes in animals. *Geochim. Cosmochim. Acta*, **42**, 495–506.

Eldridge, C.S., Compston, W., Williams, I.S., and Walshe, J.L. (1989) Sulfur isotopic analyses on the SHRIMP ion microprobe. In *New Frontiers in Stable Isotope Research*, Shanks, W.C. and Criss, R.E., eds. U.S. Geological Survey Bulletin, Vol. 1890, pp. 163–174.

Epstein, S. and Zeiri, L. (1988) Oxygen and carbon isotopic compositions of gases respired by humans. *Proc. Natl. Acad. Sci. U.S.A.*, **85**, 1727–1731 (March 1988, Geophysics).

Faure, G. (1986) *Principles of Isotope Geology*, 2nd ed. John Wiley & Sons, New York.

Friedman, I. and O'Neil, J.R. (1977) Compilation of stable isotope fractionation factors of geochemical interest. In *Data of Geochemistry*, Fleischer, M., ed. U.S. Geological Survey Professional Paper 440 KK.

Hoefs, J. (1987) *Stable Isotope Geochemistry*. Springer-Verlag, New York.

Keeling, C.D., Bacastow, R.B., Carter, A.F., Paper, S.C., Whorf, T.P., Heimann, M., Mook, W.G., and Roeloffzen, H. (1989) A three-dimensional model of atmospheric CO_2 transport based on observed winds: 1. Analysis of observational data. *Am. Geophys. Union, Geophys. Mon.*, **55**, 165–236.

Kirshenbaum, I. (1951) In *Physical Properties and Analysis of Heavy Water*, Urey, H.C. and Murphy, G.M., eds. McGraw-Hill, New York.

Lide, D.R. (1991) *Handbook of Chemistry and Physics*, 71st ed. CRC Press, Boston, Massachusetts.

Nier, A.O. (1947) A mass spectrometer for isotope and gas analysis. *Rev. Sci. Instrum.*, **18**, 398–411.

Nier, A.O. (1989) Isotope analyses with small-volume mass spectrometers. In *New Frontiers in Stable Isotope Research*, Shanks, W.C. and Criss, R.E., eds. U.S. Geological Survey Bulletin, Vol. 1890, pp. 1–8.

O'Neil, J.R. (1986a) Theoretical and experimental aspects of isotopic fractionation. In *Stable Isotopes in High Temperature Geological Processes*, Valley, J.W., O'Neil, J.R., and Taylor, H.P., Jr., eds. Mineralogical Society of America, Reviews of Mineralogy, Vol. 16, pp. 1–40.

O'Neil, J.R. (1986b) Terminology and standards. In *Stable Isotopes in High Temperature Geological Processes*, Valley, J.W., O'Neil, J.R., and Taylor, H.P., Jr., eds. Mineralogical Society of America, Reviews of Mineralogy, Vol. 16, pp. 561–570.

Pupezin, J., Jakli, G., Jansco, G., and Van Hook, W.A. (1972) The vapor pressure isotope effect in aqueous systems. I. H_2O-D_2O ($-64°$ to $100°$) and $H_2^{16}O$-$H_2^{18}O$ ($-17°$ to $16°$); ice and liquid. II. Alkali metal chloride solution in H_2O and D_2O (-5 to 100). *J. Phys. Chem.*, **76**, 743–762.

Suess, H.E. (1987) *Chemistry of the Solar System: An Elementary Introduction to Cosmochemistry*. Wiley, New York.

Walker, F.W., Parrington, J.R., and Feiner, F. (1989) *Nuclides and Isotopes*, 14th ed. General Electric Co., San Jose, California.

Zinner, E. (1989) Isotope measurements with the ion microprobe. In *New Frontiers in Stable Isotope Research*, Shanks, W.C. and Criss, R.E., eds. U.S. Geological Survey Bulletin, Vol. 1890, pp. 145–162.

2

Isotopic Exchange and Equilibrium Fractionation

2.1 Isotopic Exchange Reactions

Equilibrium isotopic fractionations are best understood in terms of reactions that involve the transfer of isotopes between two phases or molecular species that have a common element (M). These isotopic exchange reactions may be written in one of several standard forms, such as

$$aAM_b^* + cBM_d = aAM_b + cBM_d^*$$

(2.1)

where AM_b and BM_d represent the chemical formulas of the phases or species, AM_b^* and BM_d^* represent the same phases or species in which the trace isotope has replaced some or all of the atoms of element M, and a, b, c, and d are stoichiometric coefficients. In the case where all of the molecules are homogeneous, that is, where AM_b and BM_d are composed solely of the common isotope of M, and where AM_b^* and BM_d^* are phases or species in which the trace isotope M^* has replaced all atoms of element M, then the product $a \times b$ equals $c \times d$ and represents the total number of atoms exchanged in the reaction.

The concept of the isotopic exchange reaction is best shown by an example. Consider the exchange of deuterium between water and hydrogen gas. This may be written as a reaction among isotopically homogeneous molecules; that is,

$$H_2O + D_2 = D_2O + H_2$$

(2.2a)

40

or, alternatively, as exchange between homogeneous and heterogeneous molecules:

$$H_2O + HD = HDO + H_2 \tag{2.2b}$$

Much of the utility of isotopic exchange reactions is that they may be described by equilibrium constants, defined in the standard way as the quotient of the activities of the products and reactants. Thus, the equilibrium condition for equation 2.2b becomes

$$K = \frac{[HDO][H_2]}{[H_2O][HD]} \tag{2.3}$$

where K is the equilibrium constant. In equation 2.3, K has a particularly high value of 3.7 at 25°C.

With these elementary definitions in place, several key questions can be addressed. These include (1) What are the magnitudes of the equilibrium constants for isotopic exchange reactions, and what are the governing variables? (2) Can ordinary thermodynamic relationships be used to evaluate K in isotopic exchange reactions, and, if not, how can K be evaluated? (3) What is the relationship between the isotope ratio R and the thermodynamic activities of the different isotopic species? (4) What is the relationship between K and the isotopic fractionation factor α? (5) What is the dependence of α and K on temperature and pressure?

2.2 Basic Equations

2.2.1 Classical Thermodynamics

Classical thermodynamics embodies an elegant set of fundamental relationships among mathematically conservative functions and variables, such as temperature and pressure, that characterize the state of a chemical system. In most cases, the properties of isotopic molecules ("isotopomers") are so similar that standard calorimetric measurements are not sufficiently precise to make accurate predictions of their relative behavior. However, the thermodynamic differences among molecules that contain different isotopes of the lightest elements, particularly hydrogen, are sometimes sufficiently large that useful predictions can be made with classical techniques.

Of particular utility here is the fact that mixtures of compounds and their isotopomers are ideal solutions. Very effective use can therefore be made of several well-known, classical relationships between the equilibrium constant (K) and other state variables. The best known of these is

$$\boxed{\Delta G^0_{rxn} = -\boldsymbol{R}T \ln K} \tag{2.4}$$

which provides the quantitative relationship between K, absolute temperature T (kelvins), the gas constant $\boldsymbol{R}$ (denoted in bold to distinguish it from the isotope ratio R), and the standard Gibbs free energy change (ΔG_{rxn}^0) for the particular reaction. The latter quantity is defined when all the reactants and products are in the standard state, generally defined as unit concentrations at 1 atm and the temperature of interest. The latter phrase, "the temperature of interest," is key here, and proper use of equation 2.4 requires that a value of ΔG_{rxn}^0 be provided for each temperature for which $\ln K$ is determined. As a result, equation 2.4 is sometimes referred to as the "reaction isotherm." This caveat warns against a very common misuse of this relationship, notably to infer an incorrect relationship between $\ln K$ and $1/T$ that denies the strong dependence of ΔG_{rxn}^0 on the temperature.

Another powerful classical relationship is

$$\frac{\partial \ln K}{\partial T} = \frac{\Delta H_{rxn}^0}{RT^2} \qquad (2.5a)$$

which expresses the relationship between the equilibrium constant and the enthalpy change (ΔH_{rxn}^0) associated with a reaction. If the functional relationship between ΔH_{rxn}^0 and temperature can be determined, which is commonly possible because the heat capacities on which the enthalpies depend are often tabulated as power series of the temperature, then equation 2.5a can be integrated. Alternatively, over limited temperature ranges ΔH_{rxn}^0 can be assumed to be approximately constant, and equation 2.5a may be directly integrated:

$$\ln K = -\frac{\Delta H_{rxn}^0}{RT} + \text{constant} \qquad (2.5b)$$

An important application of equation 2.5b is the "Clausius-Clapeyron" equation, which relates the vapor pressure P of a substance to the temperature T and its heat of vaporization (ΔH_{vap}^0). Thus,

$$\ln \frac{P}{P_0} = -\frac{\Delta H_{vap}^0}{R} \left(\frac{1}{T} - \frac{1}{T_0} \right) \qquad (2.5c)$$

where P_0 refers to the known vapor pressure at a reference temperature T_0.

2.2.1.1 Example Calculations   The data in Table 2.1 can be used to determine the equilibrium constant for the exchange of deuterium between liquid water and hydrogen gas (equation 2.2a). Finding by subtraction that the value for ΔG_{rxn}^0 is -780 cal/mol, and using the value of 1.987 cal/mol-deg for the gas constant R, equation 2.4 may be used to calculate a value for K of 3.73, in agreement with the measured value of 3.7.

A more interesting example is the exchange of D between liquid and gaseous water molecules. The exchange equation may be written as

$$(H_2O)_{liq} + (HDO)_g = (HDO)_{liq} + (H_2O)_g \qquad (2.6)$$

Table 2.1 Thermodynamic Data for some Hydrogen Compounds at 25°C (Lide, 1991)

Substance	ΔH_f^0 (kcal/mol)	S^0 (cal/mol-deg)	ΔG_f^0 (kcal/mol)
H_2 (g)	0.00	31.208	0.00
HD (g)	0.076	34.343	−0.350
D_2 (g)	0.00	34.620	0.00
H_2O (g)	−57.796	45.104	−54.634
H_2O (liq)	−68.315	16.71	−56.687
HDO (g)	−58.628	47.658	−55.719
HDO (liq)	−69.285	18.95	−57.817
D_2O (g)	−59.560	47.378	−56.059
D_2O (liq)	−70.411	18.15	−58.195

Again, we can use the data in Table 2.1 to determine that the value for ΔG_{rxn}^0 is −45 cal/mol for this exchange reaction. Equation 2.4 may then be used to calculate a value of 1.079 for the equilibrium constant, which is in good agreement with laboratory measurements.

A key point can now be made that relates to the highly similar properties of compounds and their isotopomers. The values for ΔG_{rxn}^0 for isotopic exchange reactions are clearly determined by the small differences between similar numbers. The net free energy differences are typically only on the order of a few calories. These differences are very small compared with the values for ΔG_{rxn}^0 of ordinary chemical reactions, which typically are several kilocalories. Thus, even though the free energies of formation are known for every participating species in equation 2.2b to very high accuracy, only two significant figures remain after their subtraction, and the resulting value of −45 cal for ΔG_{rxn}^0 is very small. Thus, although there is a distinct preference for the deuterium to concentrate in the liquid water relative to the vapor, this preference is weak, and K is accordingly close to unity. Isotope exchange reactions for elements other than hydrogen generally have even smaller preferences to fractionate the isotopes, so the values for ΔG_{rxn}^0 are even smaller than that for the last example; of course, the values of K are even closer to unity. For these elements, thermochemical data can generally not be measured to the degree of accuracy needed to make useful predictions, and other means are used to estimate isotopic fractionations.

2.2.1.2 Vapor Pressures   The data in Table 2.1 may also be used to determine the vapor pressures of pure H_2O, pure HDO, and pure D_2O liquids at 25°C. This is done by first determining the values for ΔG_{rxn}^0 for the three following reactions:

$$H_2O_{liq} = H_2O_{vapor} \tag{2.7a}$$
$$HDO_{liq} = HDO_{vapor} \tag{2.7b}$$
$$D_2O_{liq} = D_2O_{vapor} \tag{2.7c}$$

These ΔG_{rxn}^0 values are 2053, 2098, and 2136 cal/mol, respectively. Assuming unit fugacity coefficients, and taking note that there are 760 torr in an atmosphere, equation 2.4 may be used to compute vapor pressures of 23.76 torr for H_2O, 22.02

torr for HDO, and 20.65 torr for D_2O. The latter values are hereafter denoted as $P^0_{H_2O}$, P^0_{HDO}, and $P^0_{D_2O}$, where the superscript 0 denotes that these vapor pressures were taken above the isotopically pure liquid materials. These values are all in good agreement with experimental determinations.

It is useful to examine now the ratios of these vapor pressures. Notably, at 25°C the numerical value for $P^0_{H_2O}/P^0_{HDO}$ is 1.079, and that for $P^0_{H_2O}/P^0_{D_2O}$ is 1.151. Note that the former number is identical to the value for the equilibrium constant for equation 2.6. It will be shown in chapter 3 that this equilibrium constant is, in fact, very important, because it is also the isotopic fractionation between liquid water and water vapor. Moreover, note that the square root of 1.151 is 1.073, a value numerically close to this fractionation factor. This similarity will be discussed further below, in section 2.6 on the "rule of the mean."

Classical thermodynamic equations may be applied to calculate many other results of interest (Figure 2.1). As an example, it might be important to know the value of the equilibrium constant for equation 2.6 at a temperature different than 25°C—say, at 50°C. This may be accomplished by first determining the heats of vaporization (ΔH^0_{vap}) for reactions 2.7a and b. The Clausius-Clapeyron equation (2.5c) may then be used to estimate that the vapor pressure increases from 23.76 torr at 25°C $(T_0 = 298.15\,K)$ to 93.83 torr at 50°C $(T = 323.15\,K)$ for ordinary H_2O, and from 22.02 to 88.56 torr for HDO over the same temperature range.

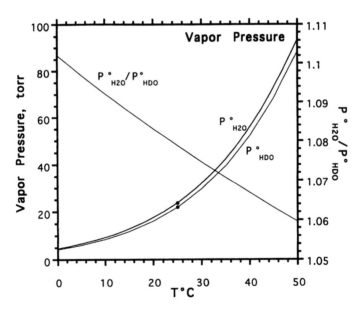

Figure 2.1 Calculated vapor pressures of pure H_2O and pure HDO (left scale), and the ratio of those vapor pressures (right scale), as a function of temperature. All information needed to generate the three curves is contained in Table 2.1, in that the vapor pressures may be calculated at 25°C (points), and, given the heats of vaporizations of the two liquids at 25°C, the Clausius-Clapeyron equation may be used to estimate the temperature dependences of these vapor pressures. The vapor pressure ratio approximates the hydrogen isotope fractionation factor between water and water vapor. See text.

The liquid–vapor fractionation factor at 50°C is then estimated to be the quotient 93.83/88.56, or 1.060, which compares reasonably well with the value of 1.0561 measured by Majoube (1971).

In short, though the contrary is commonly believed to be true, there is nothing fundamentally wrong with using classical thermodynamic methods to determine fractionation factors. The only problem is that the enthalpies and free energies are generally not known to sufficient precision to permit accurate calculations to be made. The above examples are instructive exceptions to this situation, though, even in these favorable cases, much more accurate fractionations can be determined by direct laboratory measurements. Fortunately, the methods of statistical thermodynamics provide an independent theoretical means to estimate the values of fractionation factors.

2.2.1.3 Symmetry Numbers   Symmetry numbers, denoted by σ, refer to the number of indistinguishable ways that a given molecule can be oriented in space. Thus, $\sigma = 1$ for heteronuclear diatomic molecules such as HF, CO, or $^{16}O^{18}O$, but $\sigma = 2$ for homonuclear diatomic molecules such as H_2 or $^{16}O^{16}O$. More complicated molecules can have higher numbers; for example, $\sigma = 12$ for CH_4. Symmetry numbers are important parameters in several subsequent equations.

2.2.1.4 Relationship between Species Activities and Isotope Ratios   Many important links between isotope distribution and classical thermodynamics can be made if a relationship between the isotope ratios of a substance and the activities of its various isotopomers can be established. Consider an element with only two isotopes, M and M^*, whose relative abundances are denoted by x and y. Then,

$$x + y = 1 \tag{2.8a}$$

If the two isotopes are randomly distributed among s identical sites in a molecule, the classical relative abundances of each species can be determined by individual terms of the binomial expansion:

$$(x + y)^s = x^s + sx^{s-1}y + (s/2)(s - 1)x^{s-2}y^2 + \cdots + y^s \tag{2.8b}$$

Because of equation 2.8a, all these relative abundances must sum to unity; that is, to the sth power of $x + y$. In the ideal case, the individual terms will conform to the relative activities of the various molecular species. The normalized activity of any given species is then given by the appropriate individual term of the binomial expansion:

$$a_{M_{s-r}M_r^*} = \frac{s!}{(s - r)!r!} x^{s-r}y^r \tag{2.8c}$$

Taking into account all the isotopomers, the overall isotope ratio R of the substance in question is clearly

$$R = y/x \tag{2.8d}$$

The relationship between R and the activities may then be determined case by case. However, a useful generalization between R and the activity ratio, a^*/a, of any two given isotopomers is (R. E. C., unpublished)

$$R = \left(\frac{a^*\sigma^*}{a\sigma}\right)^{1/b}$$ (2.9)

where b is the difference between the number of sites occupied by the heavy isotope in the two isotopomers. This relationship appears to be valid for a large variety of simple molecules, and it is consistent with the Ferronsky and Polyakov rule given later (equation 2.20). Complications arise for molecules that have distinguishable varieties for single isotopic compositions. Examples include left- and right-handed varieties, or the trans and cis forms of a square planar molecule. For such molecules, the relationship between R and the abundances must be established case by case. Moreover, if the different isomeric forms have different energies, then the assumption of a random distribution of isotopes will not be valid.

Similar methods may be used to determine relative abundances for cases where three or more isotopes of the element of interest are present. For example, for three isotopes with abundances x, y, and z in a molecule with s sites, the relative abundances of the various compositional species are given by the individual terms of the expansion

$$1 = (x + y + z)^s$$ (2.10)

For such cases, equation 2.9 still applies, because the y/x ratio cannot be changed by the addition of the third isotope. In other words, the ratio of apples to oranges in a truck cannot be changed by the addition of apricots!

These above relationships suggest that the number of distinguishable isotopomers that relates to the distribution of p isotopes among s identical sites is given by the combinatorial equation, which becomes

$$\# \text{Species} = \frac{(p + s - 1)!}{(p - 1)!s!}$$ (2.11)

Thus, for three isotopes in a diatomic molecule of an element, six different species would be present. For a complex molecule that comprises multiple elements, the total number of species would be the product of the individual terms of the form of equation 2.11—one term for each type of site. Of course, this type of calculation will not include isomeric forms such as cis and trans varieties.

2.2.2 Statistical Thermodynamics

Classical thermodynamics investigates the interrelationships between the macroscopic properties of chemical systems and the governing system variables, with the point of view that these properties can be *measured*. In contrast, statistical thermodynamics uses the microscopic properties of systems and statistical theory to *calculate* these macroscopic properties of chemical systems. This calculation is accomplished by investigating the possible microscopic states of a system, and then computing the overall value of a given macroscopic property by taking a suitable statistical average of the values exhibited by all the constituent microscopic states. In principle, all of the information needed to calculate practically any thermodynamic property is contained within the spectroscopic properties of a substance, because the observed frequencies are intimately related to particular energy states that have discrete, quantized levels.

It is instructive to examine a few fundamental relationships in this field. Suppose we wish to characterize the energy relationships in a system that comprises N particles that have several allowed energy levels. That is, suppose that n_1 particles have energy ε_1, n_2 have energy ε_2, and so on. Then, the total number of particles N is simply

$$N = \sum_j n_j \tag{2.12a}$$

while the total internal energy of the system is

$$E = \sum_j n_j \varepsilon_j \tag{2.12b}$$

The average energy of the particles is obtained by dividing the total energy E by the total number of particles N:

$$\bar{E} = \frac{\sum_j n_j \varepsilon_j}{N} = \sum_j P_j \varepsilon_j \tag{2.13}$$

where the probability P_j is the fraction of molecules that have energy ε_j. The problem of defining thermodynamic parameters with statistical theory reduces to defining these probabilities, which can be accomplished with the Boltzmann distribution law. According to this law, the fraction or probability P_i of molecules that will have energy ε_j decreases exponentially with the magnitude of that energy level. Specifically,

$$P_j = \frac{n_j}{N} = \frac{g_j e^{-\varepsilon_j/kT}}{\sum_j g_j e^{-\varepsilon_j/kT}} \tag{2.14}$$

where k is Boltzmann's constant (1.38×10^{-23} J/K), which is identical to the familiar "gas constant" R divided by Avogadro's number. One complication is that, in certain cases, two or more distinct states might have the same energy level, and for this circumstance the term g_i, called the "degeneracy," is included in equation 2.14.

The denominator on the right-hand side of equation 2.14 is of key importance to statistical thermodynamics. This factor is given a special name, the "partition function," and is denoted by the symbol "Q":

$$Q = \sum_j g_j e^{-\varepsilon_j/kT} \qquad (2.15)$$

Equation 2.15 can be used to determine the average, macroscopic properties of a chemical system in terms of the constituent microscopic states. For example, by taking the partial derivative of Q with respect to temperature at constant N and constant volume V, it can readily be shown using equations 2.13 and 2.14 that the average internal energy $\bar{E}$ is

$$\bar{E} = kT^2(\partial \ln Q/\partial T)_{N,V} \qquad (2.16)$$

Suppose now that the chemical system of interest has different types of energy, the germane examples being translational, vibrational, rotational, and electronic energy. The total energy is clearly represented by the sum of these different contributions:

$$E_{\text{total}} = E_1 + E_2 + E_3 + \cdots E_n \qquad (2.17a)$$

Because of the logarithmic relationship between E and Q, equations 2.16 and 2.17a can be used to directly show that

$$Q_{\text{total}} = Q_1 Q_2 Q_3 \cdots Q_n \qquad (2.17b)$$

where the subscripts relate the individual Q values to the particular type of energy in equation 2.17a. Thus, the total partition function for a complex system is given by the product of the constituent partition functions, each representing a separate class of energy states.

Numerous additional quantitative relationships between Q and familiar thermodynamic functions are derived in standard books on statistical thermodynamics. For example, for a closed, isothermal system, the Helmholtz free energy "A" assumes a particularly simple form (e.g., Hill, 1960):

$$A = -kT \ln Q \qquad (2.18)$$

2.2.3 Relationship between K, Q, and α

For isotopic exchange reactions (equation 2.1), the net volume changes are negligible, because no sensible volumetric differences can arise simply by redistributing neutrons among the tiny atomic nuclei of the reactant and product atoms. Largely because of this, no PV differences arise for such equations. For this case, classical thermodynamic relationships indicate that changes in the enthalpy ΔH and internal energy ΔE are equal, given the classical relationship $H = E + PV$. Consequently, for isotopic exchange reactions, the Helmholtz free energy change, ΔA_{rxn}, is equal to the Gibbs free energy change, ΔG_{rxn}.

Given the above, the similarity between equation 2.4 and equation 2.18 suggests a close link between the partition function Q and the equilibrium constant K. Using equation 2.1 as an example, the equilibrium constant may, in fact, be written in terms of the simplified partition functions (Q values) of Urey (1947; see below), where each Q represents the equilibrium constant between the compound of interest and its separated atoms. Thus,

$$K = \left(\frac{Q_{BM_d^*}}{Q_{BM_d}}\right)^c \Big/ \left(\frac{Q_{AM_b^*}}{Q_{AM_b}}\right)^a \qquad (2.19)$$

Note that the form of this expression is the familiar one, except that the partition functions Q have replaced the activities.

The remaining task is to relate the equilibrium constant K to the isotopic fractionation factor α. In many cases (e.g., for equation 2.2b), these quantities are identical. In other cases, K and α differ by a power that relates to the stoichiometry of the reaction. It is commonly asserted that α is the nth root of K, where n refers to the number of atoms transferred in the exchange reaction. The latter rule is usually, but not always, correct. Ferronsky and Polyakov (1982) provide the following relationship, which appears to have greater validity:

$$\alpha = \left(\frac{K}{K_\infty}\right)^{1/ac} \qquad (2.20)$$

Here, the product ac is equal to the number of atoms exchanged in an appropriately written isotopic exchange reaction (e.g., equations 2.2b, 2.6, 2.22a, but not 2.2a), and K_∞ represents the limiting equilibrium constant at very high temperature. The latter term is solely determined by the stoichiometric coefficients of the exchange equation and by the "symmetry numbers" of the molecules:

$$K_\infty = \frac{(\sigma_{AM_b^*}/\sigma_{AM_b})^a}{(\sigma_{BM_d^*}/\sigma_{BM_d})^c} \qquad (2.21)$$

Note that the symmetry numbers cancel in the correct expression for α (equation 2.20), because the same factors are present in the expressions for both K and K_∞. Isotopic fractionation does not depend on classical factors such as symmetry numbers (Bigeleisen and Mayer, 1947; Stern et al., 1968).

2.2.3.1 Example   Many of the latter relationships are well illustrated by the important oxygen isotope exchange reaction between water and carbon dioxide gas:

$$2H_2{}^{18}O + C({}^{16}O)_2 = 2H_2{}^{16}O + C({}^{18}O)_2 \qquad (2.22a)$$

The equilibrium constant for equation 2.22a is written in the standard way; namely,

$$K = \frac{[H_2{}^{16}O]^2[C({}^{18}O)_2]}{[H_2{}^{18}O]^2[C({}^{16}O)_2]} \qquad (2.22b)$$

The isotopic fractionation factor (α) for oxygen may be directly written as

$$\alpha = \frac{R_{CO_2}}{R_{H_2O}} = \frac{({}^{18}O/{}^{16}O)_{CO_2}}{({}^{18}O/{}^{16}O)_{H_2O}} \qquad (2.22c)$$

The relationship between α and K for the equilibrium exchange between CO_2 and H_2O may be found if, among the isotopomers constituting each phase, the oxygen isotopes are distributed randomly. Suppose then that the abundance of ${}^{18}O$ in the H_2O is represented by f and that the abundance of ${}^{16}O$ in the H_2O is represented by g. It is immediately clear that the concentration of $H_2{}^{18}O$ is f, that the concentration of $H_2{}^{16}O$ is g, and therefore that the ${}^{18}O/{}^{16}O$ ratio of the H_2O is simply the ratio f/g. Addition of any amount of $H_2{}^{17}O$ to the system, by the way, cannot affect the ${}^{18}O/{}^{16}O$ ratio, so the ${}^{17}O$ isotope may be neglected in the analysis.

The situation for the CO_2 is somewhat more difficult. For this case, let the isotopic abundances of ${}^{18}O$ and ${}^{16}O$ in the CO_2 be respectively represented by x and y. Note that there are three different types of CO_2 molecules: namely, $C{}^{16}O{}^{16}O$, $C{}^{18}O{}^{18}O$, and $C{}^{16}O{}^{18}O$. The relative abundances of these species are then y^2, x^2, and $2xy$. The extra factor of 2 in the third case arises because there are two different ways to make the mixed molecule, $C{}^{16}O{}^{18}O$ and $C{}^{18}O{}^{16}O$, and this fact doubles the likelihood of the occurrence of this molecule. Note that the probabilities for observing these CO_2 molecules are calculated in a similar way as would be used to predict the outcomes of two consecutive coin tosses, except that x and y take on the appropriate values of the abundances for the oxygen isotopes rather than both being equal to 0.5.

The ${}^{18}O/{}^{16}O$ ratio for the CO_2 may now be defined by examining the abundances of the individual isotopomers. Specifically, this ratio must equal the number of ${}^{18}O$ atoms in the CO_2 divided by the number of ${}^{16}O$ atoms in the CO_2:

$$\left(\frac{{}^{18}O}{{}^{16}O}\right)_{CO_2} = \frac{2[C{}^{18}O{}^{18}O]+[C{}^{16}O{}^{18}O]}{2[C{}^{16}O{}^{16}O]+[C{}^{16}O{}^{18}O]} = \frac{(2x^2+2xy)}{(2y^2+2xy)} = x/y \qquad (2.23)$$

Though the final result is obvious in retrospect (cf. equation 2.8d), the calculation requires some explanation. For the numerator, the number of ${}^{18}O$ atoms in the species $C{}^{18}O{}^{18}O$ is the abundance x^2 of this molecule, multiplied by 2 because each molecule contains two atoms of ${}^{18}O$. The number of ${}^{18}O$ atoms in the species $C{}^{18}O{}^{16}O$ is simply the abundance $2xy$ of this molecule, multiplied by unity. A similar argument can be used to compute the total number of ${}^{16}O$ atoms in the denominator. Of course, equation 2.9 could have been used to directly calculate R from the abundances of any pair of the CO_2 isotopomers.

Given that the ${}^{18}O/{}^{16}O$ ratio for the H_2O is f/g, and that for the CO_2 is x/y, then,

$$\alpha = \frac{({}^{18}O/{}^{16}O)_{CO_2}}{({}^{18}O/{}^{16}O)_{H_2O}} = \frac{(x/y)}{(f/g)} \qquad (2.24)$$

All that remains to be done is to evaluate K, which can be accomplished by substituting the above-determined abundances of the various isotopomers into equation 2.22b for the equilibrium constant K. Thus,

$$K = \frac{(x^2/y^2)}{(f/g)^2} = \alpha^2 \qquad (2.25)$$

Clearly, in this case, α is simply equal to the square root of K. This result is consistent with the rule of Ferronsky and Polyakov (equation 2.20). First, the number of atoms exchanged in the reaction (equation 2.22a) as written is 2, which corresponds to the exponent of this relationship. Moreover, the symmetry number of every participating molecule in the reaction is equal to 2. Because of this, the value for K_∞ determined from equation 2.21 is unity, and so, in this case, this factor provides no complication.

2.3 Molecular Models

The previous equations provide the basis for the theory of isotopic fractionation. It is necessary to first determine the partition function Q for each of the various participating species in any isotope exchange reaction of interest. Actually, only the ratio of the Q values for a given molecule and its isotopomer is needed, because this information is sufficient to determine K (equation 2.19), and then α can be determined from K using the Ferronsky and Polyakov rule (equation 2.20).

As shown in equation 2.14, calculation of Q requires that all the energies of all the microscopic states are known. Although this requirement might appear to be unattainable, the following sections illustrate how this can be accomplished through the use of molecular models whose parameters can be determined from spectroscopic data. It will become apparent that the effects of equilibrium isotopic fractionation primarily arise from the differences in the vibrational frequencies of different materials.

2.3.1 Simple Harmonic Oscillator

It is useful to examine the vibrational behavior of the simplest relevant chemical species, the diatomic molecule. A surprisingly useful approximation to this behavior can be given by making an analogy to the familiar physical entity of two masses connected by a spring, known as the "simple harmonic oscillator," or SHO (Figure 2.2a). The chemical bond between the molecules is assumed to act like an "ideal" spring, for which the force exerted is proportional to the amount of stretching. That is,

$$F = -kx \qquad (2.26a)$$

where x is the displacement from the neutral position and k is the "spring constant." The potential energy V is found by integrating this force over the distance displaced, so the energy well is seen to be parabolic; that is,

Figure 2.2 Simple physical models for harmonic oscillators composed of point masses connected by a spring. (a) Case where two masses undergo simple harmonic oscillations whose frequency depends on the spring constant and on the reduced mass of the system, according to the equations given in the text. (b) Case where one mass is infinite, for which the vibrational frequency depends only on the spring constant and on the mass of the object in motion. See text.

a.

b.

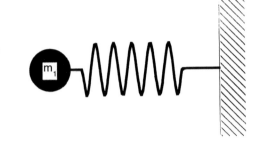

$$V = kx^2/2 \tag{2.26b}$$

Now, a differential equation can be written from the force equation (2.26a), given that the acceleration (a) is simply the second derivative of displacement with time. First, consider the limiting case where a weight of mass m is connected by the spring to an infinitely heavy mass (Figure 2.2b). The force equation becomes

$$F = ma = m\,d^2x/dt^2 = -kx \tag{2.27a}$$

which has the solution

$$x = A\cos\{(k/m)^{1/2}t\} \tag{2.27b}$$

The reader can immediately verify these results, and can show by differentiation that equation 2.27b satisfies equation 2.27a.

For the more germane case where two finite masses, m_1 and m_2, are connected by a spring, the equation is slightly more complex, but is treated in many elementary physics books. In short, the equations may be greatly simplified through use of the reduced mass μ, defined as

$$\mu = \frac{m_1 m_2}{m_1 + m_2} \tag{2.28}$$

The force equation becomes

$$\mu\,d^2x/dt^2 = -kx \tag{2.29a}$$

which is the same as equation 2.27a except that μ has replaced m. The solution is

$$x = A\cos\{(k/\mu)^{1/2}t\} \tag{2.29b}$$

This result has the same form as equation 2.27b, and, in fact, reduces to that equation for the case where m_2 approaches infinity. Note that for one cycle of motion, the argument of the cosine in equation 2.29b becomes

$$(k/\mu)^{1/2}T = 2\pi \qquad (2.29c)$$

where T is the period. Rewriting the last result in terms of the frequency ν which is simply the inverse of the period, gives

$$\boxed{\nu = \frac{1}{2\Pi}\sqrt{\frac{k}{\mu}}} \qquad (2.29d)$$

Equation 2.29d is very useful because it directly expresses the vibrational frequency of the two-body system in terms of the reduced mass μ and the spring constant k. Of particular importance here, this equation may be used to compare the vibrational frequency of a diatomic molecule with that of its isotopomer. In that case, while the masses change because of the addition of one or more neutrons, the chemical bond, which arises from interactions in the electronic shell, will be unaffected. The spring constant can be eliminated by writing two equations of the form of 2.29d, one for the molecule and one for the isotopomer, and dividing to obtain

$$\boxed{\frac{\nu_1}{\nu_2} = \sqrt{\frac{\mu_2}{\mu_1}}} \qquad (2.30)$$

This identical ratio also applies if the frequencies are expressed in wavenumbers (ω). The latter are commonly used to report data, and are calculated by simply dividing the frequencies by the speed of light c; that is, $\omega = \nu/c$.

2.3.1.1 Example

The utility of equation 2.30 is best shown by example. The diatomic molecule $^{16}O - {}^{16}O$ has a single vibrational frequency of 1580.193 cm^{-1} (Table 2.2; Huber and Herzberg, 1979). From this result, we could use equations 2.28 and 2.30 to predict that the isotopomer $^{16}O - {}^{18}O$ would have a

Table 2.2 Spectroscopic Data of Some Diatomic Molecules (Huber and Herzberg, 1979)

Gas	μ (amu)	ω_e (cm^{-1})	$\omega_e\chi_e$ (cm^{-1})	B_e (cm^{-1})	D_0 (cm^{-1})	R_e (Å)
H_2	0.50391261	4401.213	121.336	60.853	36118.26	0.74144
HD	0.67171137	3813.15	91.65	45.655	36406.20	0.74142
D_2	1.00705111	3115.50	61.82	30.4436	36748.91	0.74152
HF	0.95705545	4138.32	89.88	20.9557	47336.30	0.916808
DF	1.8210454	2998.192	45.761	11.0102	47892.82	0.91694
$H^{35}Cl$	0.97959272	2990.9463	52.8186	10.5934	35759.11	1.274552
$D^{35}Cl$	1.90441364	2145.163	27.1825	5.448794	36175.29	1.274581
$^{16}O_2$	7.99745751	1580.193	11.981	1.44563	41259.77	1.20752

smaller frequency of 1535.57 cm^{-1}. For simple systems, the frequency shifts associated with isotopic substitutions can be accurately estimated and understood. However, such simple arguments will generally not work for complex systems, especially solids and liquids.

2.3.1.2 SHO Vibrational Energies   In order to compute Q, it is necessary to determine and then sum up the energy levels associated with the molecule. For vibrational energies, this has been accomplished by solving the Schrodinger equation to obtain the energy levels of a molecule that is undergoing simple harmonic oscillations. While this demonstration is outside the scope of this book, this well-known result of quantum mechanics is elegant and compact:

$$\varepsilon_{SHO} = (1/2 + n)h\nu$$

(2.31a)

where ε_{SHO} is the vibrational energy, n is the energy level (which is an integer: 0, 1, 2, …), h is Planck's constant, and ν is the frequency. This result predicts that an infinite number of equally spaced energy levels arise for every fundamental frequency of vibration (Figure 2.3). A key feature of this equation is that the

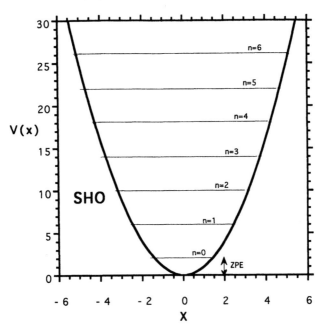

Figure 2.3 Quantized energy levels for a simple harmonic oscillator (SHO). The energy well for the SHO is parabolic, as defined by the bold curve of energy $V(x)$ as a function of the displacement x. The units of V and x are arbitrary, except V is here defined as zero at the equilibrium position ($x = 0$) defined by the bottom of the well. The various energy levels, enumerated from $n = 0$ to $n = 6$, and so on, are discrete and equally spaced in this quantized system. The energy difference between the lowest level ($n = 0$) and the bottom of the well is the zero point energy, or ZPE.

lowest vibrational level ($n = 0$) does not have zero energy, but rather has the value $h\nu/2$. This value is known as the "zero point energy," or ZPE:

$$\boxed{\text{ZPE} = h\nu/2}$$

$$(2.31\text{b})$$

This energy exists even at a temperature of absolute zero (0 K), and it is arguably the single most important factor that leads to the phenomenon of equilibrium isotope fractionation.

Given this detailed knowledge of the energy levels associated with the vibrational frequency, and the theory that the population of the various energy levels will follow a Boltzmann distribution, equation 2.15 can be used to compute Q. Thus, for a simple harmonic oscillator,

$$Q_{\text{SHO}} = \sum_i e^{-\varepsilon_i/kT} = \sum_{n=0}^{\infty} e^{-(n+1/2)h\nu/kT} \qquad (2.32\text{a})$$

This result may be simplified to

$$Q_{\text{SHO}} = e^{-U/2} \sum_{n=0}^{\infty} e^{-nU} \qquad (2.32\text{b})$$

where the term $e^{-U/2}$ has been factored out of the expression, and where U is defined as

$$U = \frac{h\nu}{kT} = \frac{hc\omega}{kT} \qquad (2.32\text{c})$$

Finally, the series can be written in closed form:

$$Q_{\text{SHO}} = \frac{e^{-U/2}}{1 - e^{-U}} \qquad (2.32\text{d})$$

because of the well-known mathematical relationship

$$\sum_{n=0}^{\infty} x^n = \frac{1}{1 - x} \qquad \text{for } 0 < x < 1 \qquad (2.32\text{e})$$

The SHO partition function is then seen as the product of two parts, one of which arises from the zero point energy:

$$\boxed{Q_{\text{ZPE}} = e^{-U/2}}$$

$$(2.33\text{a})$$

and the other of which relates to the vibrational energy spacings of the SHO:

$$\boxed{Q_{\text{vib}} = \frac{1}{1 - e^{-U}}}$$

$$(2.33\text{b})$$

2.3.1.3 SHO Rotational Energies   The contribution to Q that arises from rotational energy of the diatomic molecule can be similarly evaluated by molecular models. In the most straightforward case, the diatomic molecule can be modeled as a rigid, rotating "dumbbell." Again, the Schrodinger equation can be solved to obtain the quantized energy levels of such a "molecule," yielding another well-known result:

$$\varepsilon_{rot} = hcB_eJ(J+1) \tag{2.34}$$

where ε_{rot} is the rotational energy, J is the energy level (which is an integer: $0, 1, 2, \ldots$), h is Planck's constant, c is the speed of light, and B_e is a constant. Again, this result predicts that an infinite number of quantized, rotational energy levels arise for every type of rotation. The resultant partition function is approximately

$$\boxed{Q_{rot} = \frac{8\pi^2 IkT}{\sigma h^2}} \tag{2.35}$$

where I is the moment of inertia of the molecule.

2.3.2 Morse Potential

In several respects, the SHO model is unrealistic for molecules. The potential energy well associated with the SHO is a simple parabola (Figure 2.3), and within it are contained an infinite set of equally spaced, quantized energy levels. No matter how much energy is added, no provision can be made for the dissociation of the molecule into its constituent atoms, nor for the enormous increase in repulsion if the molecule is compressed too greatly.

Several more realistic potential functions have been devised to describe molecules, but the simplest of these is the empirical Morse potential (Morse, 1929; Herzberg, 1950). According to this model, the potential energy $V(r)$ is

$$\boxed{V_{(r)} = D_e[(1 - e^{-\beta(r-r_e)})^2 - 1]} \tag{2.36a}$$

where r is the interatomic distance, r_e is the equilibrium distance, and D_e is the depth of the energy well. Also, β is a constant that is given numerically in cm^{-1} by the relation (Huber and Herzberg, 1979)

$$\beta_{cm^{-1}} = 1.2176 \times 10^7 \omega_e \sqrt{\mu/D_e} \tag{2.36b}$$

where μ is in amu, and D_e and ω_e are in cm^{-1}. Last, the depth of the energy well, D_e, is equal to the sum of the zero point energy, given in equation 2.38a, and the dissociation energy D_0 of the molecule in the ground state:

$$D_e = D_0 + ZPE \tag{2.36c}$$

Note that if D_0 and D_e are given in cm^{-1}, they must be multiplied by the speed of light c and by Planck's constant h to be converted into energy units; alternatively, they may be multiplied by the numerical constant 2.85848 to convert to cal/mol.

Many of the aforementioned spectroscopic parameters are given in Table 2.2. As shown in Figure 2.4, the Morse function is asymmetrical, so that as r approaches infinity, V approaches zero and the molecule will dissociate. Moreover, when r equals r_e, V is equal to $-D_e$, and it can be shown by differentiation that V is a minimum at this point. Last, the value of V also increases very rapidly as the interatomic distance r approaches zero, although it does not approach infinity as would be required for a real molecule. Nevertheless, the Morse potential provides a much better model for a molecule than does the SHO.

By solving the Schrodinger equation, Morse (1929) rigorously showed that the combined rotational and vibrational energy levels associated with his potential are

$$E_{(n,J)}/hc = \omega_e(n + \tfrac{1}{2}) - \omega_e\chi_e(n + \tfrac{1}{2})^2 + B_eJ(J + 1) + \cdots \qquad (2.37)$$

where $E_{(n,J)}$ is the energy associated with the vibrational level n and the rotational level J, h is Planck's constant, c is the speed of light, ω_e is the fundamental mode of vibration of the molecule in wavenumbers (cm^{-1}), and $\omega_e\chi_e$ and B_e are con-

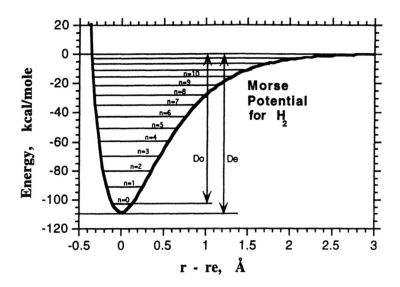

Figure 2.4 Quantized energy levels for the diatomic hydrogen molecule, as approximated by a Morse oscillator. The energy well for hydrogen is not a simple parabola, but is close to the Morse potential (bold curve), and has the finite depth D_e. If the interatomic distance r becomes too large, so that the molecule is stretched too far from the equilibrium position r_e, which happens to be 0.74144 Å for H_2, then the molecule will dissociate. The dissociation energy D_0 represents the depth to the lowest energy level, $n = 0$, while the zero point energy is the difference between D_e and D_0. Note that the quantized energy levels are not uniformly spaced in this case, because the Morse potential has an anharmonic character. The entire graph was generated with equations given in the text and the H_2 data provided in Table 2.2.

stants that may be derived from spectroscopic data. The first and third terms on the right-hand side are identical to the vibrational and rotational energy levels associated with a rigid, rotating molecule that is undergoing harmonic oscillations (equations 2.31a, 2.34).

The second term in equation 2.37 is known as the "anharmonicity correction" to the first term. Because this term is negative, and because it increases rapidly as the quantum level n increases, then the vibrational levels are not equally spaced, but rather become progressively closer (Figure 2.4).

For cases where the "anharmonicity correction" is significant, it has two consequences. First, the ZPE is modified to become

$$\text{ZPE} = hc\{\omega_e/2 - \omega_e\chi_e/4\} \tag{2.38a}$$

Consequently, the partition function associated with the ZPE becomes

$$Q_{\text{ZPE}} = \exp\left[-\frac{hc}{kT}\left\{\frac{\omega_e}{2} - \frac{\omega_e\chi_e}{4}\right\}\right] \tag{2.38b}$$

Second, this new term also gives rise to a new partition function for anharmonicity, which Richet et al. (1977) suggest can be expressed as

$$Q_{\text{anh}} = 1 - 2\chi_e U e^{-U}/(1 - e^{-U})^2 \tag{2.38c}$$

This factor must be multiplied by the partition function Q_{vib} for the SHO in order to obtain the corrected vibrational partition function. Alternatively, the correction may be obtained by simply summing up the appropriate vibrational terms in the series in equation 2.15. Such anharmonicity factors will not be used further in this treatment, but, where they are needed, they may easily be included. Richet et al. (1977) discuss these and other terms that may be used to refine calculations of partition functions.

2.3.3 Solids

The spectroscopic characteristics of solids are complex, so the partition functions are necessarily complex as well. For a unit cell composed of N atoms, there are $3N$ degrees of freedom in the solid. Consequently, there are $3N$ vibrational modes, because, of course, in this case no rotational or translational modes are possible. Three of these modes are low-energy acoustic modes that relate to sound velocity; most of the others correspond to lattice vibrations and bends. In only a few of these cases will the frequency shifts attendant upon isotopic substitution be given simply by an SHO model. Kieffer (1982) provides a model for the estimation of frequency shifts associated with isotopic substitution in solids, along with some calculations of the oxygen isotope fractionation factors of silicate minerals. Spectroscopic measurements of solids that comprise solely heavy isotopes are needed to refine such models, but very few data of this type are available.

2.4 Theory of Isotopic Fractionation

2.4.1 Qualitative Explanation of Isotopic Fractionation

The material presented previously provides a sufficient basis for the development of a theory of isotopic fractionation. Before proceeding with Urey's quantitative theory, it is useful to develop some qualitative concepts. Most important is a "rule of thumb":

> The heavy isotope concentrates in the compound in which the element is bound most strongly.

A straightforward explanation for this generality is evident in the previous equations. The energy associated with a given vibrational mode is directly proportional to the frequency of that mode (equation 2.31a). These vibrational frequencies strongly depend on the "force constant" of the bond, and on the masses of the atoms in motion. Equations 2.29d and 2.30 also show that the vibrational frequency of an SHO will be lowered by substituting a heavy isotope for a light isotope, due to the increase in the reduced mass. Such substitutions will not affect the bond strength, because the latter arises from interactions in the electronic shell, and these are the same for all isotopes of a single element in a given chemical environment.

This explanation is still incomplete. It is clear that any substance will have an affinity to incorporate the heavy isotope, because the energy level of the substance will invariably be lowered by this isotopic substitution. The question is: Among two or more different substances, which will preferentially incorporate the heavy isotope?

In fact, the heavy isotope will be most concentrated in the substance that has the highest force constant, which typically, but not always, is the substance with the highest frequencies. Such a "high-energy" compound will "win out" in this competition, because the difference in the energy levels between such a compound and its isotopomer will be greater than this same difference will be between other compounds and their isotopomers (see example in section 2.4.1.1).

Several corollaries of the above generalities can be made. For example, bonds to ions with high ionic potential and low atomic mass tend to incorporate the heavy isotope preferentially, because those factors are related to the bond strength. Consider the partitioning of ^{18}O between various oxides, particularly those of silicon and iron. In oxides, silicon is small (ionic radius $= 0.41$ Å) and highly charged ($+4$) compared to iron (ionic radius $= 0.76$ Å; charge $= +2$), so the heavy isotope would be expected to concentrate along with the silicon. In fact, among the common rock-forming minerals, quartz (SiO_2) is observed to have the greatest tendency to concentrate ^{18}O, while iron minerals like magnetite (Fe_3O_4) have some of the lowest observed ^{18}O concentrations. As yet another example, sphalerite (ZnS) concentrates ^{34}S relative to galena (PbS).

The difference in the energy levels also explains nonequilibrium fractionations that arise from the "kinetic isotope effect." These quantum effects arise from the difference in the molecular dissociation energies that relate to the levels of the various ZPEs (Figure 2.5). In particular, because their ZPEs are lower, molecules

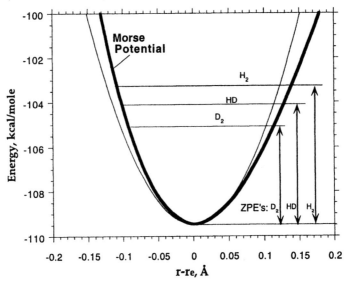

Figure 2.5 Detail of the base of the Morse potential (bold curve) shown in Figure 2.4, whose asymmetric character is evident when compared with a simple parabola (light curve). The zero point energies for H_2, HD, and D_2, representing the energy difference between the bottom of the well and the ground state ($n = 0$ levels) for the three molecules, were calculated from the data in Table 2.2. The differences among the ZPEs are the fundamental cause of equilibrium isotope fractionation. These differences also give rise to differences in dissociation energies for the three molecules, which can produce kinetic isotope effects.

containing heavy isotopes sit lower in the potential energy well than do molecules with light isotopes, and more energy is accordingly required to break their bonds. Because the isotopically light molecules are more easily dissociated, these molecules preferentially enter into chemical reactions. According to Bigeleisen (1965), up to tenfold reductions in reaction rates have been realized by the substitution of a heavy isotope for a lighter one in critical molecular locations.

2.4.1.1 Example   Consider the partitioning of deuterium between two very similar substances, hydrogen chloride and hydrogen fluoride. At low temperatures, practically all of the atoms will be in the ground state, so the overall vibrational energy of the substances is well represented by the appropriate ZPEs. The relative sense of the ZPEs in these compounds is indicated by the magnitudes of the fundamental vibrational frequencies, because it is directly proportional (equation 2.31a). The vibrational frequency of HF is 4138.32 cm^{-1}, so, using equation 2.30, the frequency of DF can be estimated to be 2998.19 cm^{-1}. Similarly, the fundamental frequency for $H^{35}Cl$ is 2990.946 cm^{-1}, so the frequency of $D^{35}Cl$ can be estimated to be 2145.16 cm^{-1}. Thus, the relative ZPEs can be shown schematically as in Figure 2.6.

It is evident that the frequency lowering associated with isotopic substitution is greatest in the hydrogen fluoride; this is a plausible result given that fluorine is the

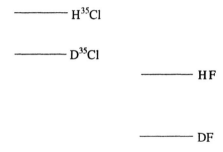

Figure 2.6 Diagram of the relative positions of the zero point energies ($n = 0$ levels) for HF, HD, $H^{35}Cl$, and $D^{35}Cl$. In a system composed of these molecules, deuterium will preferentially concentrate in the hydrogen fluoride, because the ZPE difference between HF and DF is greater than that between the hydrogen chlorides.

most electronegative element that exists in nature. Fluorine also has a substantially smaller ionic radius (1.36 Å) than does Cl (1.81 Å). In order to complete the analysis, it is convenient to view this situation in terms of an isotopic exchange reaction:

$$HF + DCl = DF + HCl \tag{2.39}$$

As argued by Bigeleisen (1965), the energy change for such a reaction is given to first order by the differences among the ZPEs, which is

$$\Delta E^0{}_{ZPE} = (ZPE)_{DF} + (ZPE)_{HCl} - (ZPE)_{DCl} - (ZPE)_{HF} \tag{2.40}$$

It is evident that the reduction in energy of the system is greatest if most of the deuterium is partitioned into the hydrogen fluoride, insofar as the difference between the ZPEs of HF and DF is greater than the difference between HCl and DCl . It is this relative energy change, rather than the magnitude of any particular energy level, that favors the production of the compounds on the right-hand side of equation 2.39.

2.4.2 Urey's Theory

The statistical mechanical theory of Urey (1947) defines the partition function ratios of gaseous molecules and their isotopomers. Urey "simplified" the equations to directly compare each compound with its separated atoms. The condition of separated atoms serves as a very useful "baseline" for examining isotopic exchange reactions, because the number of any given type of atom must be the same on both the right-hand side and the left-hand side of an exchange equation. In the final analysis, where the equilibrium constants and fractionation factors are determined, the question becomes, given all these constituent atoms and isotopes: Do energy considerations favor the chemical products on the right-hand side or the left-hand side of the exchange equation, and by exactly how much?

Urey's "simplified" ratio of the partition functions for a diatomic molecule (subscript 1) and its isotopomer (subscript 2) becomes the product of several different contributions (cf. equation 2.17b):

$$\frac{Q_2}{Q_1} = \frac{\sigma_1}{\sigma_2} \frac{U_2}{U_1} \quad \frac{e^{-U_2/2}}{e^{-U_1/2}} \quad \frac{1 - e^{-U_1}}{1 - e^{-U_2}} \tag{2.41}$$

Rotation   ZPE   Vibrational energy spacings

where the U_i represent the quantities $h\nu_i/kT$, and the σ_i are the symmetry numbers, as before. The first term principally relates to the rotational behavior and symmetry of the molecule, and was basically derived from equation 2.35 using a spectroscopic theorem of Teller and Redlick (Urey, 1947). The other two factors in the expression can be understood by reference to equations 2.33a,b. Thus, if the single vibrational frequency of the diatomic molecule and that of its isotopomer are both known from spectroscopic study, or if the latter can be estimated from the former using the SHO model (equation 2.30), then the Q_2/Q_1 ratio for the pair can be directly determined.

Similar results can be obtained for more complex molecules, but these have multiple frequencies that must all be taken into account. Nonlinear polyatomic molecules have $3N - 6$ vibrational modes, where N is the number of atoms (e.g., 3 for H_2O), representing $3N$ degrees of freedom less 3 translational modes and 3 rotational modes. Following equation 2.17b, the resulting ratio of simplified partition functions for these nonlinear molecules must be a product term:

$$\frac{Q_2}{Q_1} = \frac{\sigma_1}{\sigma_2} \prod_{i=1}^{3n-6} \frac{U_{2i}}{U_{1i}} \frac{e^{-U_{2i}/2}}{1 - e^{-U_{2i}}} \frac{1 - e^{-U_{1i}}}{e^{-U_{1i}/2}} \tag{2.42}$$

The expression for linear polyatomic molecules is the same as the latter, except that there are $3N - 5$ vibrational modes instead of $3N - 6$. These modes represent $3N$ degrees of freedom less 3 translational modes and less 2 rotational modes. This difference of 1 rotational mode arises because a linear molecule such a CO_2 cannot store any rotational energy by spinning around its long axis, so there is no sensible frequency associated with any such motion.

2.5 Temperature Dependence of Isotopic Fractionation Factors

2.5.1 Theoretical Results

One of the most important and useful characteristics of isotopic fractionation is that it depends only on the temperature, and not on any other variable of state. The lack of dependence on pressure arises because no sensible changes in volume are associated with isotopic exchange reactions such as equation 2.1, principally because the atomic nuclei that host the neutrons constitute only about one-trillionth of the total volume of the atoms. Possible exceptions may ultimately be found in certain mineral–water or solid–gas equilibria (see O'Neil, 1986). In this book, isotopic exchange reactions are viewed as perfect geothermometers, and the manner in which these equilibria vary with temperature is of great interest.

Much insight into the dependence of isotopic fractionation factors on temperature can be gleaned by examination of Urey's equations, discussed in the previous section. In particular, the functional behavior of these basic equations can be mathematically examined in three limiting cases: notably (1) where the U_i are large, which occurs at low temperature, and/or if the vibrational frequencies are high; (2) where the U_i are small, representing the condition at high temperatures, and/or for low frequencies; and (3) where the U_i are effectively zero, which occurs as infinite temperature is approached. These conditions are referred to in the following simply as the "low temperature," "high temperature," and "infinite temperature" cases, respectively.

2.5.1.1 Low-Temperature Case   At low temperatures (large U), it is evident that the $(1 - e^{-U})$ factors all approach unity. Equation 2.41 for the simplified partition function ratios for diatomic molecules reduces to

$$\frac{Q_2}{Q_1} \cong \frac{\sigma_1}{\sigma_2} \frac{U_2}{U_1} e^{(U_1 - U_2)/2} \tag{2.43a}$$

Similarly, for nonlinear polyatomic molecules, equation 2.42 simplifies to

$$\frac{Q_2}{Q_1} \cong \frac{\sigma_1}{\sigma_2} \prod_{i=1}^{3n-6} \frac{U_{2i}}{U_{1i}} e^{(U_{1i} - U_{2i})/2} \tag{2.43b}$$

The logarithm of these ratios may then be written as

$$\ln \left[\frac{Q_2}{Q_1} \right] \cong \frac{U_1 - U_2}{2} + \ln \left[\frac{\sigma_1 v_2}{\sigma_2 v_1} \right] \tag{2.44a}$$

for diatomic molecules, and

$$\ln \left[\frac{Q_2}{Q_1} \right] \cong \sum_{i=1}^{3n-6} \frac{U_{1i} - U_{2i}}{2} + \sum_{i=1}^{3n-6} \left[\ln \frac{\sigma_1}{\sigma_2} \frac{v_{2i}}{v_{1i}} \right] \tag{2.44b}$$

for nonlinear polyatomic molecules, respectively (Criss, 1991). Note that equations 2.44a and b are linear in $1/T$, being of the form $y = m/T + \text{constant}$, where the slope m is the difference between the zero point energies $(hv_i/2)$ of the ordinary molecule and its isotopomer, divided by Boltzmann's constant. The dependence of $\ln(Q_2/Q_1)$ on the zero point energy difference arises because, at low temperatures, the $n = 0$ vibrational energy levels are populated by practically all of the molecules.

2.5.1.2 High-Temperature Case   The behavior of the partition function ratios at high temperatures is best learned by expanding Urey's equations in a Taylor series. This can be readily accomplished because, at high temperatures, the U_i become very small. The results are simplest if the logarithms are first taken of equations 2.41 and 2.42, and, although the development of such series is tedious, derivations are given by Bigeleisen (1958) and Criss (1991). The series representation of equation 2.41 for diatomic gases is

$$\ln\left[\frac{Q_2}{Q_1}\right] = \ln\left[\frac{\sigma_1}{\sigma_2}\right] + \frac{U_1^2 - U_2^2}{24} - \frac{U_1^4 - U_2^4}{2880} + \frac{U_1^6 - U_2^6}{181,440} - \frac{U_1^8 - U_2^8}{9,676,800} + \cdots$$

(2.45a)

Similarly, equation 2.42 for polyatomic molecules becomes

$$\ln\left[\frac{Q_2}{Q_1}\right] = \ln\left[\frac{\sigma_1}{\sigma_2}\right] + \sum_{i=1}^{3n-6}\left\{\frac{U_{1i}^2 - U_{2i}^2}{24} - \frac{U_{1i}^4 - U_{2i}^4}{2880} + \frac{U_{1i}^6 - U_{2i}^6}{181,440} - \frac{U_{1i}^8 - U_{2i}^8}{9,676,800} + \cdots\right\}$$

(2.45b)

The $U^2/24$ dependence of $\ln[Q_2/Q_1]$ was originally deduced by Urey (1947), and is evident from the second term of the series in equations 2.45a and b. The higher the temperature, the less important the higher order terms in the series become. Thus,

$$\ln\left[\frac{\sigma_2 Q_2}{\sigma_1 Q_1}\right] \cong \frac{U_1^2 - U_2^2}{24}$$

(2.46a)

for diatomic molecules, and

$$\ln\left[\frac{\sigma_2 Q_2}{\sigma_1 Q_1}\right] \cong \sum_{i=1}^{3n-6}\frac{U_{1i}^2 - U_{2i}^2}{24}$$

(2.46b)

for nonlinear polyatomic molecules. Appendix A.3 provides some simple equations for the differences between the squared frequencies of normal and isotopically substituted molecules.

2.5.1.3 Infinite-Temperature Limit

Last, it is useful to consider this behavior in the limiting case where temperature approaches infinity. In this limit, equations 2.45a and b both reduce to

$$\frac{Q_2}{Q_1} = \frac{\sigma_1}{\sigma_2}$$

(2.47)

This last result, which along with equation 2.19 leads directly to the classical value for K_∞ (equation 2.21), can also be obtained by applying l'Hopital's rule to either of equations 2.41 and 2.42—again, as $T \to \infty$. Of course, because K and K_∞ are identical in this high-temperature limit, then, according to the Ferronsky and Polyakov rule (equation 2.20), the isotopic fractionation factor α must approach unity. Thus, as $T \to \infty$, all of the following conditions apply:

$$K \to K_\infty \tag{2.48a}$$

$$\alpha^{eq} \to 1 \tag{2.48b}$$

$$1000 \ln \alpha^{eq} \to 0 \tag{2.48c}$$

$$\Delta^{eq} \to 0 \tag{2.48d}$$

2.5.1.4 Temperature Dependence of Isotopic Fractionation Factors Equations 2.19 and 2.20 show that the equilibrium constant K and the isotopic fractionation factor α are related to the Q_2/Q_1 ratios by stoichiometric powers. This means that $\ln K$, $\ln \alpha$ and $\ln Q_2/Q_1$ must be similarly related by simple multiplicative constants that depend on these same stoichiometric coefficients. It follows that $\ln K$ and $\ln \alpha$ will vary as functions of $1/T$ at low temperatures, as is generally understood. In fact, using equations 2.19 and 2.21, it is straightforward to rewrite equation 2.20 as follows

$$\ln \alpha = \frac{1}{a} \ln \left[\frac{\sigma_{BM_d^*} Q_{BM_d^*}}{\sigma_{BM_d} Q_{BM_d}} \right] - \frac{1}{c} \ln \left[\frac{\sigma_{AM_b^*} Q_{AM_b^*}}{\sigma_{AM_b} Q_{AM_b}} \right] \tag{2.49}$$

Making appropriate substitutions (from equation 2.44a) for the Q_2/Q_1 ratios of the diatomic molecules A and B at low temperatures gives

$$\ln \alpha \cong \frac{c(U_{BM_d} - U_{BM_d^*}) - a(U_{AM_b} - U_{AM_d^*})}{2ac} + \ln \left[\left(\frac{\nu_{BM_d^*}}{\nu_{BM_d}} \right)^{1/a} \left(\frac{\nu_{AM_b}}{\nu_{AM_b^*}} \right)^{1/c} \right] \tag{2.50}$$

Note that the symmetry numbers have canceled. More important, this expression is linear in $1/T$, since the first term on the right is proportional to U. Of course, this expression is only valid at low temperatures for which the approximations on which it rests are valid. For example, because of the constant term on the right, it is impossible for this equation to reduce to the necessary condition that α approaches unity—that is, $\ln \alpha$ approaches zero—as temperatures approach infinity. Instead, equation 2.40 defines a finite temperature where $\alpha = 1$, a condition known as a "crossover" (see section 2.5.2.4).

Proceeding similarly (i.e., by combining equations 2.49 and 2.46a) for exchange reactions that involve diatomic molecules at elevated temperatures gives

$$\ln \alpha \cong \frac{c\left(U_{BM_d}^2 - U_{BM_d^*}^2 \right) - a\left(U_{AM_b^2} - U_{AM_b^*}^2 \right)}{24ac} \tag{2.51}$$

This expression is linear in $1/T^2$, and, because it has an intercept of zero, it extrapolates to the necessary condition that α approaches unity as temperature approaches infinity. Once again, the symmetry numbers have no influence on the magnitude of the isotopic fractionation factors.

In addition to representing these well-known temperature proportionalities, equations 2.50 and 2.51 explicitly indicate the theoretical values for the slopes and y-intercepts of the fractionation lines in the low- and high-temperature cases.

These two equations apply only to isotopic exchange between diatomic molecules; equations for exchange reactions that involve polyatomic molecules are similar in form but contain sum and product terms. A linear, zero intercept relation between $\ln \alpha$ and $1/T^2$ also appears to have useful application to the empirical description of oxygen isotopic equilibria among nonhydrous silicates or oxides. In effect, this occurs because the values of U are generally small for these substances at geologically relevant temperatures (Bottinga and Javoy, 1975).

It needs to be pointed out, however, that equations 2.50 and 2.51 cannot predict many of the complex behaviors that may be exhibited in gaseous equilibria, and which are accounted for by Urey's original equations, 2.41 and 2.42. For example, equation 2.51 does not predict any crossovers (see section 2.5.2.4), inflection points, or maxima or minima in fractionation equations. Equation 2.50 predicts a single crossover, albeit one that may lie outside its range of validity (see section 2.5.2.4). An interesting discussion and examples of the complex effects that can occur in natural systems is given by Stern et al. (1968).

Equations of the form $\ln \alpha = C_1 + C_2/T + C_3/T^2$, where the C values are empirical constants, are commonly used to describe laboratory determinations of certain fractionation factors that, over temperature ranges of interest, do not appear to strictly follow either the $1/T$ or the $1/T^2$ limiting dependencies. For a single phase, no such combination of terms can describe the partition function ratios, except in the unlikely case of multiple frequencies that are so different that some are in the "small U" limit while others are in the "large U" limit at a single temperature. However, the isotopic fractionation factor makes a comparison of two different phases, which may have very different frequencies. This situation might occur if the fractionation of interest involves distinctly different materials, such as a gas and a liquid or solid. In such a case, if the frequencies are greatly different, the first term of equation 2.49 could take on the $1/T$ form suggested by equation 2.44b, and the second term could take on the $1/T^2$ form suggested by equation 2.46b. In fact, Jones (1958) argues theoretically that equations of this form (with $C_1 = 0$) are useful in the description of the vapor pressure ratios of isotopic solids.

Of course, the range of equation 2.51 can be extended by inclusion of additional terms; for example,

$$\ln \alpha \cong \frac{c\left(U_{BM_d}^2 - U_{BM_d^*}^2\right) - a\left(U_{AM_b}^2 - U_{AM_b^*}^2\right)}{24ac}$$
$$+ \frac{c\left(U_{BM_d}^4 - U_{BM_d^*}^4\right) - a\left(U_{AM_b}^4 - U_{AM_b^*}^4\right)}{2880ac} + \cdots \tag{2.52}$$

2.5.2 Temperature Dependence in Real Systems

Isotopic fractionation effects can be observed in many natural systems and can be studied in the laboratory. In many cases, the isotopic fractionation factors for these real systems conform to the behaviors predicted earlier, but in many other cases, the behaviors are much more complicated. The effects that have been observed include the following.

2.5.2.1 Isotopic Equilibrium   A necessary condition that isotopic equilibrium has been attained in a system is that the measured fractionation factor R_A/R_B, representing the partitioning of isotopes between two phases A and B, is equal to the equilibrium fractionation factor appropriate for the temperature of the system (equation 1.11). That is,

$$\alpha^{\text{measured}} = \alpha^{\text{eq}} \tag{2.53a}$$

2.5.2.2 Isotopic Disequilibrium   In many systems, it is observed that the measured fractionation factor R_A/R_B is not equal to the equilibrium fractionation factor appropriate for the temperature of the system. That is,

$$\alpha^{\text{measured}} \neq \alpha^{\text{eq}} \tag{2.53b}$$

In such cases, any temperature calculated to be appropriate for the observed fractionation will be different than the actual temperature. An important example, known as the "Dole effect," is that the $^{18}O/^{16}O$ ratio of atmospheric O_2 is approximately 20‰ too high to be in isotopic equilibrium with seawater. In this case, the ^{18}O content is controlled by metabolic reactions that involve photosynthesis and respiration.

2.5.2.3 Isotopic Reversal   In the case of extreme disequilibrium, the observed fractionation of isotopes may be opposite in sign to that appropriate for isotopic equilibrium. This condition is called an "isotopic reversal." No physical temperature is implied or, in some cases, can even be calculated for such systems, because the solution to a given calibration curve might give an imaginary number for the temperature.

2.5.2.4 Crossover   In many real cases, the variations of equilibrium isotopic fractionation factors with temperature are much more complicated than the simple behaviors predicted earlier for high and low temperatures. In particular, it is possible for phase A to concentrate the heavy isotope relative to phase B over one range of temperature, and for the opposite preference to occur over another temperature interval. In such a situation, the point where the isotope ratios of the two phases are equal—that is, where $\alpha^{\text{eq}} = 1$—is called a "crossover," and the temperature that corresponds to that equilibrium condition is called the "crossover temperature," T_c. Note that a crossover temperature is finite, and, as such, this phenomenon is distinguished from the fact that isotopic fractionation factors approach unity as temperature approaches infinity.

Thus, regarding the crossover phenomenon, there are three conditions:

$$
\begin{array}{ll}
\text{for } T < T_c & \alpha^{\text{eq}} > 1 \text{ (or vice versa)} \\
\text{for } T = T_c \ll \infty & \alpha^{\text{eq}} = 1 \\
\text{for } T > T_c & \alpha^{\text{eq}} < 1 \text{ (or vice versa)}
\end{array} \tag{2.54}
$$

A fine example of a crossover is exhibited by the fractionation of hydrogen isotopes between water and water vapor (Figure 2.7). It is seen that, at temperatures below 224°C, the liquid water concentrates deuterium relative to the vapor, so $\alpha^{eq} > 1$. As temperatures climb to the crossover temperature of 224°C, the D/H ratios of the liquid and vapor are identical, so $\alpha^{eq} = 1$ and, therefore, $\ln \alpha^{eq} = 0$. At still higher temperatures, above the crossover, the vapor now has a higher D/H ratio than the coexisting liquid, so $\alpha^{eq} < 1$. At even higher temperatures, the fractionation factor progressively approaches unity, ultimately to become unity at and above the critical point of water.

In one case, the simplified equations derived in the previous section can be used to predict a crossover point. Specifically, a crossover temperature can be calculated from equation 2.50 by setting $\ln \alpha^{eq} = 0$. In any given chemical system, this calculated condition may or may not be real, because it may or may not lie outside the range of validity of the "low-temperature" condition used to derive equation 2.50 in the first place. An example of an artificial "crossover point" is given in Criss (1991).

Again, the fractionation behavior of real isotopic systems may be extremely complex. It is even possible for the fractionations in natural systems to exhibit local maxima, minima, and multiple crossover points (see Stern et al., 1968).

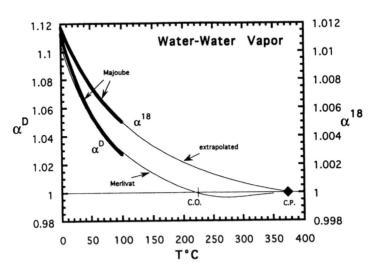

Figure 2.7 Fractionation factors between water and water vapor for hydrogen (left scale) and oxygen (right scale) isotopes, as experimentally determined by Majoube (1971; heavy curves) and Merlivat et al. (1963; light curves). If extrapolated, Majoube's curve for oxygen isotope fractionation attains a value of unity at the critical point of water (C.P., 374°C), fulfilling a thermodynamic requirement. The hydrogen isotope fractionation factor likewise becomes unity at the critical point, and also exhibits a "crossover" (C.O.) close to 224°C.

2.5.3 Natural Systems

In light of the preceding discussion, it is useful to inquire as to how commonly isotopic equilibrium is attained in natural systems. Figure 2.8 illustrates some oxygen isotope fractionation factors that involve important interactions between ocean water and either the marine atmosphere or calcium carbonate precipitated from it. Natural data from the marine-atmosphere system commonly show a close approach to the isotopic fractionation effects predicted by the curves. For example, many organisms precipitate calcareous shell materials in equilibrium with seawater at ambient temperatures, although others incorporate metabolic or kinetic effects. Similarly, atmospheric water vapor can be close to isotopic equilibrium with the ocean water, but, more commonly, an additional kinetic fractionation is present because the humidity is less than 100%, or the air and sea temperatures are not identical. Last, while atmospheric carbon dioxide is generally close to isotopic equilibrium with ocean water, the $\delta^{18}O$ value of atmospheric oxygen (O_2) differs from the predicted equilibrium value by approximately 20 per mil. In the first case, equilibria that involve dissolved bicarbonate ion provide a rapid mechanism for isotopic exchange between carbon dioxide and water, while no such mechanism exists for oxygen gas.

2.5.3.1 Representation of Real Systems Laboratory calibrations of isotopic fractionation factors have been made for numerous real systems; the results have been compiled by Friedman and O'Neil (1977). These experiments are necessarily conducted over a limited range of temperatures, and in several instances isotopic

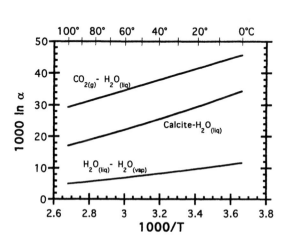

Figure 2.8 Oxygen isotope fractionation factors between water and water vapor, carbon dioxide and water, and calcite ($CaCO_3$) and water as a function of temperature; data from Table 2.3 and Appendix A.4. If all the phases were in isotopic equilibrium at 25°C ($1000/T = 3.354$) with ocean water with $\delta^{18}O = 0$, then the water vapor would be predicted to have a value of -9.3, the carbon dioxide would be $+41.3$, and the calcite would be $+28.8$ per mil. The latter values are reasonably close to those observed in the lower marine atmosphere and in recent shell materials. In detail, kinetic factors can also play a role in the actual systems. Modified after Criss (1995).

equilibrium was not attained, in part due to the short time frames available in the laboratory. Considerable effort has been directed toward conducting definitive experiments and toward confirming the results with different experimental approaches (e.g., O'Neil, 1986).

The types of observed fractionation behaviors vary from simple to very complex. Over the range of temperatures investigated, some systems appear to closely follow either the $1/T$ and $1/T^2$ types of dependencies predicted by the "low-temperature" and "high-temperature" approximations to Urey's equations (e.g., equations 2.44 and 2.46). In other instances, equations of the form $\ln \alpha = C_1 + C_2/T + C_3/T^2$, where the C values are empirical constants, are used to describe the fractionation factors. In all cases, the coefficients are empirically derived, and the equations are valid only over the ranges of temperature where the defining experiments were conducted.

In short, experimental data for fractionation factors may be recast in one of the following four forms:

$$(1) \quad 1000 \ln \alpha^{eq} = C_1 + C_2/T \qquad \text{Low } T \qquad (2.55a)$$

$$(2) \quad 1000 \ln \alpha^{eq} = C_1 + C_3/T^2 \qquad \text{High } T \qquad (2.55b)$$

$$(3) \quad 1000 \ln \alpha^{eq} = C_3/T^2 + C_4/T^4 \qquad (2.55c)$$

$$(4) \quad 1000 \ln \alpha^{eq} = C_1 + C_2/T + C_3/T^2 \qquad (2.55d)$$

Form 1 (equation 2.55a) is useful for many systems at low temperatures; for example, for oxygen isotope partitioning between carbon dioxide gas and water at low temperatures.

Form 2 (equation 2.55b) is very useful for equilibria among solids, which tend to have low vibrational frequencies so that the "high-temperature" condition is effectively attained. For most solid–solid reactions, as is clearly shown for oxygen isotope partitioning among many silicates, $C_1 = 0$ in this equation, although $C_1 \neq 0$ for solid–water equilibria or for equilibria involving hydrous phases (e.g., Bottinga and Javoy, 1973). Only in the case where $C_1 = 0$ can this expression converge to the required condition where α approaches unity as temperature approaches infinity, illustrating the deficiencies of this type of empirical expression.

Form 3 (equation 2.55c) has been used only recently, but equation 2.52 suggests that it may have some utility at temperatures that are not high enough for the $1/T^2$ proportionality to be strictly accurate.

Form 4 (equation 2.55d) encompasses both of the above types of behaviors, and is useful for describing complex equilibria over limited temperature ranges. In part, this utility may simply arise because there are more free parameters that can be used to obtain a fit to the experimental data. However, as discussed above, this form may be especially useful for solid–gas equilibria.

Table 2.3 provides some examples of important isotopic fractionation factors; additional examples are given in Table 3.2 and Appendix A.4. Many additional determinations are given in the compilation of Friedman and O'Neil (1977) and in Richet et al. (1977). O'Neil (1986) provides numerous other data sources.

Table 2.3 Selected Isotopic Fractionation Factors

$$\text{Form: } 1000 \ln \alpha^{eq}_{A-B} = C_1 + 10^3 C_2/T + 10^6 C_3/T^2$$

Type	Phase A	Phase B	C_1	C_2	C_3	T (K) range	Ref.
D/H	Water	Water vapor	52.612	−76.248	24.844	273–373	1
$^{13}C/^{12}C$	HCO_3^-	CO_2 (g)	−24.10	9.552	0	278–398	2
$^{13}C/^{12}C$	Calcite	CO_2 (g)	2.4612	−7.6663	2.988	273–773	3
$^{18}O/^{16}O$	CO_2 (g)	Water	−15.3	16.60	0	273–373	4
$^{18}O/^{16}O$	Calcite	Water	−2.89	0	2.78	273–773	5
$^{18}O/^{16}O$	Quartz	Magnetite	0	0	6.29	873–1573	6
$^{34}S/^{32}S$	Sphalerite	Galena	0	0	0.80	523–837	7

References: (1) Majoube, 1971; (2) Mook et al., 1974; (3) Bottinga, 1968 (calculation); (4) see Appendix A.4; (5) O'Neil et al., 1969; (6) Chiba et al., 1989; (7) Kajiwara and Krouse, 1971.

2.5.3.2 Combined Results for Real Systems   If two independent fractionation equations contain one phase in common, it is commonly useful to combine these equations into a new equation that directly relates the other two phases. For example, consider the definitions of the following fractionation factors:

$$\alpha_{A-B} = \frac{R_A}{R_B} \qquad \alpha_{C-B} = \frac{R_C}{R_B} \qquad \alpha_{A-C} = \frac{R_A}{R_C} \qquad (2.56)$$

It is evident that

$$\alpha_{A-C} = \frac{\alpha_{A-B}}{\alpha_{C-B}} \qquad (2.57)$$

For equations that represent the logarithm of α—for example, equations 2.55a–d—this latter computation is effected by simply subtracting the analogous numerical coefficients. However, fractionation equations thus derived will have greater uncertainty than the original determinations; moreover, their range of validity may be very restricted, making extrapolations dangerous. For example, combination of the quartz–water fractionation with either the albite–water or the magnetite–water fractionations in Appendix A.4 gives geologically unreasonable quartz–albite and quartz–magnetite fractionation equations that disagree with the new determinations by Chiba et al. (1989).

2.6 Rule of the Mean

Another very useful law of isotope distribution, called the "rule of the mean," can be understood using the equations at hand. This rule describes relationships between compounds that have more than one isotopomer. According to Bigeleisen (1955), this rule is exact if the heat of mixing of compounds and their isotopomers is zero (i.e., $\Delta H_{mixing} = 0$), a necessary condition of ideal mixing. This rule is also exact in the case where the partition functions obey the $U^2/24$ law (equations 2.46a and b). Several forms of this rule—one for isotopic disproportionation reactions, one for vapor pressures among isotopomers, and one for fractionations of an element with multiple isotopes—are discussed in the following sections.

2.6.1 Isotopic Disproportionation Reactions

Isotopic disproportionation reactions involve the transfer of isotopes between a homogeneous molecule and its homogeneous isotopomer, to make an isotopically heterogeneous molecule. Such transfers may be written as disproportionation reactions:

$$(b - a)AM_b + aAM_b^* = bAM_{b-a}M_a^* \tag{2.58}$$

According to the "rule of the mean," this disproportionation reaction has the equilibrium constant

$$\boxed{K = \frac{(\sigma_{AM_b})^{b-a}(\sigma_{AM_b^*})^{a}}{(\sigma_{AM_{b-a}M_a^*})^{b}}} \tag{2.59}$$

In effect, for the stated conditions, K takes on the classical value given by the quotient of the symmetry numbers raised to the appropriate powers, analogous to the situation for K_∞ (equation 2.21).

The rule of the mean can be readily confirmed for the simple case of a diatomic molecule, MM, and its two isotopomers, MM^* and M^*M^*. The disproportionation reaction becomes

$$MM + M^*M^* = 2MM^* \tag{2.60}$$

The equilibrium constant K for this reaction may be directly expressed in terms of the appropriate partition functions:

$$K = \frac{(Q_{MM^*})^2}{(Q_{MM})^1(Q_{M^*M^*})^1} \tag{2.61}$$

Assuming the SHO model, the relative frequencies of all these molecules may be related by manipulating the relationship between molecular frequencies and their reduced masses (equation 2.30), to obtain

$$v_{MM^*}^2 = \frac{v_{MM}^2 + v_{M^*M^*}^2}{2} \tag{2.62}$$

Given this relationship between the frequencies, for high temperatures where the $U^2/24$ proportionality holds (equation 2.46), a general relationship between all of the relevant symmetry numbers and partition functions directly follows:

$$\sqrt{\frac{\sigma_{M^*M^*}Q_{M^*M^*}}{\sigma_{MM}Q_{MM}}} = \frac{\sigma_{MM^*}Q_{MM^*}}{\sigma_{MM}Q_{MM}} \tag{2.63}$$

Upon substituting this last result into equation 2.61, it is seen that the equilibrium constant for the disproportionation reaction reduces to a quantity dependent solely on the symmetry numbers. Again, the symmetry numbers are equal to 2 for the homogeneous diatomic molecules and to 1 for the heterogeneous diatomic molecule. Thus, for reactions such as equation 2.60,

$$K = \frac{(\sigma_{MM})^1 (\sigma_{M^*M^*})^1}{(\sigma_{MM^*})^2} = 4 \qquad (2.64)$$

It is useful to examine how closely the "rule of the mean" is satisfied at ordinary temperatures. Urey (1947) calculated the equilibrium constants for several different disproportionation reactions (Table 2.4). These constants are a function of temperature, typically varying from a value of zero at 0 K to the value of 4 at sufficiently high temperatures. For reactions involving carbon dioxide, diatomic oxygen, and diatomic nitrogen molecules at 273.1 K, the values for K closely approximate the theoretical value of 4. However, at room temperature significant deviations from the rule of the mean are seen for the reactions that involve hydrogen compounds. Moreover, these calculations are more difficult to perform accurately, because corrections for anharmonicity and other effects are necessary. The calculation for the complex water molecule is the most difficult, and Urey's (1947) calculated value of 3.96 at 25°C differs significantly from the value of 3.76 determined by Kirshenbaum (1951), from the value of 3.76 ± 0.02 measured by Friedman and Shiner (1966), from our classical thermodynamic calculation of 3.52 given in section 2.6.1.1, and, of course, from the value of 4 indicated by the rule of the mean.

The rule of the mean has an important consequence for isotopic exchange reactions. In cases where the rule is very closely satisfied, then for the calculation of α it makes little difference whether the exchange reaction is written in terms of isotopically homogeneous molecules or in terms of isotopically heterogeneous molecules. However, consider the situation where the rule of the mean is not satisfied. In this case, the Ferronsky and Polyakov rule (equation 2.20) will similarly not be satisfied. As amplified in the following, this situation principally arises for the exchange reactions that involve hydrogen. As a specific example, if α were calculated for hydrogen isotope exchange among the homogeneous molecules in equation 2.2a, the result would differ from that calculated for equation 2.2b, which is written in terms of heterogeneous molecules. The correct result for α will be obtained only in latter case, where the properties of the heterogeneous molecules are specifically used in the computation.

Table 2.4 Equilibrium Constants for Some Isotopic Disproportionation Reactions (Urey, 1947)

Reaction	$K_{0°C}$	$K_{25°C}$
$C^{16}O_2 + C^{18}O_2 = 2C^{18}O^{16}O$	3.9990	
$^{14}N_2 + {}^{15}N_2 = 2{}^{14}N^{15}N$	3.9918	
$^{16}O_2 + {}^{18}O_2 = 2{}^{16}O^{18}O$	3.9859	
$H_2 + D_2 = 2HD$	3.18	3.25
$H_2O + D_2O = 2HDO$	3.94*	3.96*

*Inaccurate.

2.6.1.1 Examples   An important example of an isotopic disproportionation reaction is

$$H_2O + D_2O = 2HDO \tag{2.65}$$

A classical thermodynamic equation (equation 2.4) and the data in Table 2.1 can be used to calculate the equilibrium constant for this reaction. If all the water molecules are in the gas phase, then the free energy change at 25°C is −745 cal/mol, yielding a value of 3.52 for K. A slightly different result (3.56) would be obtained for the reaction in the liquid phase. The discrepancy of this result from the value 4 indicates that, in this case, the rule of the mean is not strictly satisfied.

As another example, consider the disproportion reaction

$$C^{16}O_2 + C^{18}O_2 = 2C^{16}O^{18}O \tag{2.66}$$

The equilibrium constant may be calculated if it is assumed that the abundances of the various isotopomers conform to a random distribution. Thus, using the probabilities developed previously (see equation 2.23), K becomes

$$K = \frac{[C^{16}O^{18}O]^2}{[C^{16}O^{16}O][C^{18}O^{18}O]} = \frac{(2xy)^2}{(x^2 y^2)} = 4 \tag{2.67}$$

It is apparent that the rule of the mean is followed if there are no intermolecular fractionations, so that the isotopes are distributed in a perfectly random way.

Of course, the value of K for isotopic disproportionation reactions (equation 2.59) will not always be numerically equal to 4. Consider the reaction that involves the transfer of oxygen isotopes among the three equivalent sites in the planar carbonate ion. For this case, there are two different disproportionation reactions:

$$2C^{16}O_3 + C^{18}O_3 = 3C^{16}O_2{}^{18}O \tag{2.68a}$$

and

$$C^{16}O_3 + 2C^{18}O_3 = 3C^{18}O_2{}^{16}O \tag{2.68b}$$

In each case, a value of 27 for K would be predicted by the rule of the mean, because σ is equal to 6 for both of the homogeneous ions, but only equal to 2 for the heterogeneous ions on the right-hand side.

The same value can be determined by assuming that the oxygen isotopes are distributed among these CO_3 ions in a perfectly random way. If the abundances of ^{16}O and ^{18}O are respectively equal to x and y, then the probability of observing the $C^{16}O_3$ is x^3, that for $C^{18}O_3$ is y^3, that for $C^{18}O^{16}O_2$ is $3 x^2 y$, and that for $C^{16}O^{18}O_2$ is $3xy^2$. Thus, for equation 2.68a,

$$K = \frac{[C^{18}O^{16}O_2]^3}{[C^{16}O_3]^2[C^{18}O_3]^1} = \frac{(3yx^2)^3}{(x^3)^2(y^3)^1} = 27 \tag{2.69}$$

A value of 27 will also be calculated for equation 2.68b.

2.6.2 Vapor Pressure Rule

For many of the same reasons discussed above, a simple relationship obtains for the vapor pressures of a pure liquid compound and its pure liquid isotopomers. For this situation, the "rule of the mean" takes the form

$$(P^0_{AM_{b-a}M_a^*})^b = (P^0_{AM_b})^{b-a}(P^0_{AM_b^*})^a \qquad (2.70)$$

As an example of the second rule, consider the relationship among the vapor pressures of water and its deuterated isotopomers. According to equation 2.70,

$$(P^0_{HDO})^2 = (P^0_{H_2O})(P^0_{D_2O}) \qquad (2.71)$$

That this condition is approximately realized is confirmed by our previous computations for these vapor pressures at 25°C, which, again, were 23.76, 22.02, and 20.65 torr for H_2O, HDO, and H_2O, respectively. However, in detail there is a numerical discrepancy that shows imperfect adherence to the "rule of the mean." This discrepancy arises from the anharmonicity terms that have not been taken into account, and that are particularly large for molecules that contain hydrogen.

A straightforward way to view the latter relationship is to consider the plausibility that the following vapor pressure ratios are equal:

$$\frac{P^0_{HDO}}{P^0_{H_2O}} \simeq \frac{P^0_{D_2O}}{P^0_{HDO}} \qquad (2.72)$$

On each side of the equation, the principal differences between the water molecules in the numerators and those in the denominators are that the latter have one fewer deuterium atom (S. M. Savin, pers. comm., 1973).

2.6.3 Intramolecular Fractionations

Important exceptions to the rule of the mean will exist if, in a single molecule, the isotopes of a given element are not randomly distributed. This will occur if the chemical environment of the various atomic sites are not all identical. Consider the linear molecule N_2O, which is not symmetrical as one might expect, but rather has the form N—N—O. The nitrogen sites are clearly distinct, because one atom is bonded only to another nitrogen atom, while the other is bonded to both nitrogen and oxygen. In such a case, it is very likely that the $^{15}N/^{14}N$ ratio of the nitrogen atoms in one site will not be the same as the $^{15}N/^{14}N$ ratio of the atoms in the other. In such a case, an "internal" or "intramolecular fractionation" would exist.

Intramolecular fractionations may be fairly common in nature. While the three oxygen sites in the carbonate ion are all identical, those in ozone are not! Similarly, for hydrous silicate minerals, including the entire mica, clay, and amphibole mineral families, oxygen is present in both tetrahedral sites, where it is dominantly bonded to silicon, and in OH sites that clearly have distinct character. In addition to these types of oxygen, some clay minerals also contain interlayer water. Considerable analytical difficulty is associated with measuring the isotopic ratios of different sites in single substances, and few data are available.

2.6.4 Elements with Multiple Isotopes

Many elements have more than two stable isotopes, here denoted by m, m^*, and $m^\#$. The question that arises, given a fractionation factor for one pair of these isotopes, is: What would be the fractionation factor for another pair of these isotopes? As shown below, the rule of the mean may be extended to make a useful prediction of these various fractionations.

The SHO model (equation 2.29d) again may be employed to estimate the relationships among the frequencies for the various isotopes in a given molecular site. A set of equations of this form may be modified to obtain a relationship between frequencies for bonds between each of the isotopes $(m, m^*, m^\#)$ and an arbitrary mass M (R. E. C., unpublished):

$$v_{m^*}^2 - v_m^2 = z(v_{m^\#}^2 - v_m^2) \qquad (2.73a)$$

where

$$z = \left(\frac{m^\#}{m^*}\right)\left(\frac{m^* - m}{m^\# - m}\right) \qquad (2.73b)$$

Note that M has canceled out in this remarkably simple relationship. Under conditions where the $U^2/24$ rule is valid, the differences between the squares of the frequencies are directly related to the logarithms of the partition function ratios. Consequently, and again using the Ferronsky and Polyakov rule, a relationship may be obtained between the various fractionation factors among the isotopes of an element, and the parameter z (R. E. C., unpublished):

$$\alpha_{m^*/m} = (\alpha_{m^\#/m})^z \qquad (2.74)$$

As an example, for any pair of equilibrated phases, the $^{17}O/^{16}O$ fractionation is predicted to be the 0.529th root of the $^{18}O^{16}O$ fractionation.

2.7 Isotopic Thermometers

2.7.1 Carbonate Paleotemperatures

Urey (1947) generated the concept of isotopic thermometry and "paleotemperatures" in his seminal paper on isotopic fractionation. This concept has become one of the most important applications of nuclear chemistry to the earth and environmental sciences, where it is a cornerstone of paleoclimatology, paleoceanography, and isotope stratigraphy, and is widely used in studies of ore deposits and terrestrial and extraterrestrial materials.

In principle, all that is required for the determination of an isotopic temperature is that the appropriate isotopic ratios can be measured for two coexisting phases that are interrelated by a known fractionation equation, written as a function of temperature (e.g., equations 2.55). The temperature that corresponds to the measured fractionation can then be directly calculated. This temperature will be identical to the actual temperature if isotopic equilibrium between the two phases was attained.

In his conceptualization of the carbonate paleotemperature technique, Urey (1947, pp. 578–579) predicted that "a change from 0°C to 25°C should change the ^{18}O content of carbonate by 1.004 relative to liquid water and the ^{13}C content by 1.003. Accurate determinations of the ^{18}O content of carbonate rocks could be used to determine the temperature at which they were formed." Urey went on to recommend the use of oxygen in marine carbonates for paleoclimatic studies, because (1) the oceanic reservoir of oxygen is large and should have a constant isotopic concentration over time; (2) oxygen isotopes form ideal isotopic solutions for both solids and liquids; (3) oxygen isotopic compositions of carbonates may be determined with considerable precision; and (4) oxygen forms relatively stable compounds that might be expected to remain unchanged over long periods of time, thereby preserving a record of the temperature at which the carbonate formed.

The original calibration of temperature dependence of this fractionation factor was made by McCrea (1950), who prepared calcite from aqueous solutions by slow inorganic precipitation at several known temperatures. This result was slightly revised by Epstein et al. (1953), who drilled holes in mollusk shells and then analyzed the shell material that regrew under controlled conditions:

$$t°C = 16.5 - 4.3(\delta_{cc} - \delta_w) + 0.14(\delta_{cc} - \delta_w)^2 \qquad (2.75)$$

This empirical formula relates the quantity "δ_w," which actually represents the $\delta^{18}O$ value of CO_2 gas in equilibrium with the water at 25°C, with the quantity "δ_{cc}," which actually is the $\delta^{18}O$ value of the CO_2 gas that is released from the calcite when it is reacted with orthophosphoric acid (H_3PO_4) at 25°C. Note that this result is not written in the standard form for α, but rather in an "operational" format where measurements made in the laboratory can be directly introduced into the equation, with the result appearing in degrees centigrade. As such, this equation is still useful, because to this day the "CO_2 equilibration" technique of Epstein and Mayeda (1953) is routinely used to make $^{18}O/^{16}O$ determinations of water, while the orthophosphoric acid technique is still routinely used to analyze carbonate materials.

2.7.1.1 Requirements and Assumptions of the Technique   Several conditions must be satisfied in order for the carbonate paleotemperature technique to work. They may be listed as follows:

1. Precise $\delta^{18}O$ measurements of the carbonate must be made.
2. Precise $\delta^{18}O$ measurements must be made of the water in which the carbonate formed, or, alternatively, the $\delta^{18}O$ value of this water must be independently known or estimated.
3. The temperature dependence of the calcite–water fractionation factor must be known.
4. The calcite must have been formed in equilibrium with the water.
5. Subsequent to its formation, the $\delta^{18}O$ value of the calcite must not have changed.

Regarding the first three points, precise measurements of the $\delta^{18}O$ values of water and carbonate minerals are now routinely conducted in hundreds of laboratories. Moreover, the fractionation dependence of the calcite–water system is known with great precision, and, given appropriate measurements, can be used to define isotopic temperatures to within one degree of the actual value. These methods will be discussed further here.

In lieu of directly measuring the water, in paleotemperature studies the $\delta^{18}O$ value of the ocean is commonly presumed to have remained nearly constant over time. To an extent, this assumption is reasonable and very useful, because there are millions of specimens of fossil shell materials, but virtually no good samples of ancient ocean water! To first order, the isotopic composition of the ocean has probably remained nearly constant over geologic time, because its composition appears to be buffered by the vast oxygen reservoir of Earth's lithosphere (Muehlenbachs and Clayton, 1976; Gregory and Taylor, 1981). However, short-term fluctuations of ±1 per mil are known to have occurred during the Pleistocene glaciations, when sea level was lowered by ~100 m while great quantities of ice were stored on the continents. Even more problematical is the fact that the $\delta^{18}O$ value of shallow, ancient epicontinental seas, from which most of the fossil materials have been collected, may not have been typical of the modern open ocean, because they may have been modified by evaporation and by ancient meteoric runoff from the continents.

The fourth requirement, that the calcite must have been formed in equilibrium with the water, is suprisingly problematical. Numerous organisms, such as planktonic forams, brachiopods, and most mollusks, are known to precipitate their shells in isotopic equilibrium with water. In contrast, other organisms, such as calcareous algae, corals, benthic forams, and echinoderms, commonly are not in isotopic equilibrium with ambient water. Such organisms exhibit substantial "vital effects" that relate to kinetically controlled metabolic fractionations, with the disequilibrium effects commonly being greatest for the fastest growth. The complexities do not end here, because the ecological behavior of organisms may also be significant. For example, a planktonic organism might grow a calcareous test in one part of a water column, then move to another part that has a different temperature. Alternatively, an organism might principally grow its test during a certain part of the year. Seasonal isotopic variations were seen in Urey's original studies of a fossil belemnite that was carefully sectioned and analyzed. The point is that the growth patterns of an organism can greatly influence its isotopic character.

Last, the requirement that the measured $\delta^{18}O$ value of the calcite must represent its original value is commonly difficult to satisfy. Many fossil materials were originally composed of aragonite or of high-magnesium calcite, which are suscep-tible to recrystallization because they are thermodynamically unstable relative to calcite under surface conditions. Petrographic and chemical analysis, and electron microscopy, are commonly used to reveal evidence for recrystallization, over-growths, or any other effects that might indicate that the carbonate does not represent original shell material. For example, recrystallization of calcite is com-monly accompanied by a gain in trace uranium, and by losses of the minor elements Sr and Mg. Of course, if mineralogical changes or other secondary effects have occurred, the isotopic geothermometer will likely preserve only a record of subsurface conditions, typically characterized by elevated temperatures and by pore fluids that generally do not have the same $\delta^{18}O$ value as seawater. Unfortunately, only "negative" evidence can ever be used to argue that such processes have *not* disturbed a sample.

2.7.1.2 Pleistocene Climatic Oscillations   The most important result of the car-bonate paleotemperature technique has been the elucidation of the climatic changes that Earth has undergone during the last ~3 million years, a period of profound glaciation. The original work was conducted by Emiliani (1955), who made the first oxygen isotopic analyses of planktonic forams taken from deep sea cores. Emiliani found repeated $\delta^{18}O$ variations, having an amplitude of ~1.65‰, that were interpreted in terms of the repeated growth and retreat of Pleistocene ice sheets (Figure 2.9). These changes could be produced either by a ~5° temperature change, or, alternatively, by a shift in the $\delta^{18}O$ value of ocean water related to ice cap growth, but in either case, or in combination, these effects clearly relate to glacial cycles. Rather than the four glacial stages recognized by classical European

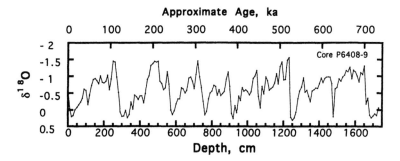

Figure 2.9 Graph of the $\delta^{18}O$ values versus depth for shells of the foram *Globigerinoides sacculifea* in a 17.4-m-long core from the Caribbean Sea, exhibiting systematic variations over time. The approximate timescale assumes a linear sedimentation rate of 2.36 cm per 1000 years to place the 700-ka Brunhes/Matuyama geomagnetic reversal at the depth of 1650 cm observed in the core, and does not take into account hiatuses and other irregula-rities in the core that are discussed by Emiliani (1978). Each of the repetitive, "sawtooth-like" patterns has a duration of about 100,000 years, and has been correlated with similar patterns at other locations in the world ocean. These patterns are the product of profound cooling and the storage of ice on continents, associated with more than a dozen major glacial advances during the Pleistocene epoch. Data from Emiliani (1978).

and American geologists, oxygen isotope studies have revealed that progressive global cooling began nearly 50 million years ago, and culminated in approximately 36 major glaciations in the last 3 Ma.

Subsequent studies of cores from throughout the world's oceans have revealed that the major climatic fluctuations are correlated. This result is significant in itself, partly because some early models predicted that glaciations would be out of phase for the northern and southern hemispheres. Moreover, the general simultaneity of the changes has afforded a new use of the isotopic records, notably isotope stratigraphy.

While the onset of major glacial periods over geologic time probably relates to the general disposition of continents over Earth's surface, the quasi-periodic climatic oscillations within a given glacial period arguably correspond to rather small variations in Earth's orbital characteristics, known as Milankovitch variations. These variations, collectively representing changes in the eccentricity, obliquity (tilt), and precession of Earth's orbit, respectively produce oscillations with periods of 100 ka, 41 ka, and 19–23 ka. Variations in eccentricity can produce up to ±5% variations in the total solar influx received by Earth, while variations in obliquity and precession redistribute the available insolation between different latitudes and different months of the year, a combination that produces climatic change.

2.7.1.3 Experimental Details   The techniques devised in Urey's laboratory for the measurement of oxygen isotopes in waters and carbonates are so robust, reliable, and rapid that they are still widely used today. These analytical techniques are so important, and are so soundly based on the principles discussed in this chapter, that it is worthwhile to include a brief discussion of them.

Because of its unreactive character, and its ease of handling under vacuum, CO_2 gas is the principal working gas in numerous mass spectrometers. As such, in order to analyze the water and carbonate, it is very useful to first convert the oxygen to this form. Different methods could be employed to achieve this for water. For example, the water could be attacked by fluorine, and the oxygen released combusted to CO_2 gas on a heated graphite rod. Instead, the much more convenient method of "CO_2 equilibration" was devised (Epstein and Mayeda, 1953).

The equilibration method exploits the observed, rapid attainment of isotopic equilibrium between water and carbon dioxide gas, which occurs by the reversible hydration of CO_2 (Mills and Urey, 1940):

$$CO_2 + H_2O = H_2CO_3 \qquad (2.76)$$

Note that the carbonic acid so formed contains a carbonate radical composed of three indistinguishable oxygen atoms, one derived from the water and two derived from the gas. A mechanism of exchange is thereby provided for the rapid isotopic equilibration of the CO_2 gas with the water. This exchange will not affect the $^{18}O/^{16}O$ ratio of the water if the ratio of the gas-to-water mole fractions is small. For this technique, then, a small amount of CO_2 gas is placed over the water sample, the system is allowed to equilibrate at a known temperature, then the CO_2 gas is collected and its $^{18}O/^{16}O$ ratio is measured with a mass spectrometer. The

$^{18}O/^{16}O$ ratio of the water can then be determined from the CO_2–H_2O fractionation factor, which is a constant at a given temperature (Table 2.3). Note that a value for this fractionation factor was not even needed to determine a paleotemperature, because of the way the original calibration equation (2.75) was set up!

A straightforward means of acidification was also devised to determine the $^{18}O/^{16}O$ ratio of carbonate from a measurement of CO_2 gas released. McCrea (1950) recognized that the acid used to react the carbonates must be anhydrous to prevent undesirable exchange between water and the CO_2 released, which occurs by the reaction just described. They accordingly prepared and utilized "100% orthophosphoric acid" for this purpose, recognizing that phosphate ion would remain intact and therefore would have no mechanism, such as reaction 2.76, to exchange with the oxygen in the carbon dioxide or calcite. Thus, the CO_2 gas released by this reaction is simply measured on the mass spectrometer, and, for the purpose of obtaining a paleotemperature by the original calibration equation, the resultant measurement is directly utilized in equation 2.75. As a bonus, if the $^{13}C/^{12}C$ ratio of the CO_2 gas is measured, it will be identical to the $^{13}C/^{12}C$ ratio of the original carbonate.

An important matter is that, while this methodology is sufficient for the paleotemperature method as originally devised, it does not in itself define the actual $^{18}O/^{16}O$ ratio of the carbonate. When carbonates are reacted with acids to release CO_2 gas, only two-thirds of the oxygen originally present in the carbonate mineral is released. Unfortunately, the $^{18}O/^{16}O$ ratio of this CO_2 gas is not equal to the $^{18}O/^{16}O$ ratio of the total carbonate. It has been determined that, at 25°C, the latter ratios are related by the "acid extraction" fractionation factor, which represents the kinetic partitioning of isotopes by this process, and not the attainment of isotopic equilibrium between calcite and the CO_2 gas. At 25°C, this factor takes on a value of 1.01025; that is,

$$\alpha_{acid}^{25°C} = \frac{(^{18}O/^{16}O)_{released\ CO_2}}{(^{18}O/^{16}O)_{calcite}} = 1.01025 \qquad (2.77)$$

In order to obtain a proper measurement of the $^{18}O/^{16}O$ ratio of a carbonate mineral, it is necessary to conduct the phosphorolysis reaction at a known temperature where this factor is known. In addition, some authors have argued that different carbonate minerals have slightly different "acid extraction" fractionation factors, but other workers use a single value, such as the calcite factor of 1.01025 given here, regardless of which carbonate mineral is being analyzed.

More recent studies have additionally disclosed that the "orthophosphoric" acid must be carefully prepared to obtain the best possible results for carbonate analysis. This acid is actually *ortho-pyro-meta*-phosphoric acid, with the formula $H_{n+2}P_nO_{3n+1}$. The polymerization of H_3PO_4 produces undesirable water by reactions such as (Wachter and Hayes, 1985)

$$2H_3PO_4 = H_4P_2O_7 + H_2O \qquad (2.78a)$$

Additional water is produced during the phosphorolysis of the carbonate; for example,

$$CaCO_3 + H_3PO_4 = CaHPO_4 + CO_2 + H_2O \qquad (2.78b)$$

As a result of the first reaction, there is an undesirable pathway to exchange oxygen in phosphate with the oxygen in the water. Water produced by either reaction can, in turn, exchange with the CO_2 gas. It therefore becomes necessary to minimize the amount of this water. Accordingly, for the best results, the acid must be very carefully prepared.

2.7.1.4 Other Carbonate Minerals   Different carbonates have somewhat different fractionation behaviors. Of the carbonate minerals in Appendix A.4, that for calcite–water is the best determined, and this calibration agrees, at low temperatures, with the paleotemperature equation. Of course, the equation used to interpret an isotopic measurement of a particular carbonate mineral must be the appropriate one.

The differences among the equations for the carbonates in Appendix A.4 are rather small, because, in this family of minerals, the oxygen is bonded more directly to the carbonate radical than it is to the divalent cation, whose influence is therefore subdued. The carbon isotope fractionations among these minerals are even more similar. Such differences are not negligible, however, and it has been argued that measurable carbon and oxygen isotope fractionations occur even between aragonite (orthorhombic $CaCO_3$) and calcite (hexagonal $CaCO_3$).

Malachite poses some additional complications, because it contains oxygen in both the carbonate and hydroxyl sites. In this case, the equation in Appendix A.4 does not represent the total isotopic fractionation between malachite and water, because the hydroxyl oxygen has not been included (Melchiorre et al., 1999). In addition, the "acid extraction" fractionation factor has not been measured because it, in any case, would not have the same meaning as it has for the other carbonates; instead, this factor has been assumed to be equal to 1.01025. For this case, the equation can be used in thermometry if the malachite is analyzed and corrected as if it were a sample of calcite.

2.7.2 Other Isotopic Thermometers

Isotopic temperatures can be calculated for any two substances that formed in, or preserve a condition of, isotopic equilibrium. While oxygen isotopes are most commonly used for this purpose, other isotopic systems can also be used. The requirements for a given isotopic thermometer to work are analogous to those stated for the paleotemperature method, in that appropriate isotopic measurements must be made, and a calibration curve must be available. Moreover, the phases must have formed in, or otherwise achieved, isotopic equilibrium, and thereafter retained those isotopic ratios. Of course, the best isotopic thermometers utilize substances that have large isotopic fractionations that depend strongly on temperature.

The temperature dependence of isotopic fractionation is so well understood that the applications are diverse. Oxygen isotope fractionations among coexisting minerals are routinely used in the study of igneous and metamorphic rocks. In many cases, slow cooling or secondary alteration processes have disturbed the original isotopic patterns, leaving records that can be used to interpret not formation temperatures, but rather the nature of the disturbance! In other cases—for example, in the study of ore deposits—if the temperature of formation is independently known or can be estimated (e.g., from fluid inclusions), the oxygen isotope ratio of a gangue mineral can be used to determine the $^{18}O/^{16}O$ ratio of the fluid that precipitated it. Sulfur isotope fractionations are also widely employed in studies of ore deposits. Hydrogen and carbon isotope fractionations are widely employed in studies of minerals, gas fields, volcanoes, and biological processes. Several of these applications will be discussed in later chapters.

The requirement of the paleotemperature method that the isotopic composition of the ocean is constant could be eliminated if a suitable mineral coprecipitated with the calcite. Much attention has turned to the phosphatic minerals, though they have proven difficult to analyze and their fractionation behavior is unfortunately similar to that of calcite. Of course, any "two-mineral" approach carries the new requirement that the materials must have originally achieved, and thereafter retained, mutual isotopic equilibrium.

2.8 Summary

Equilibrium isotopic fractionations may be understood in terms of isotopic exchange reactions that represent the transfer of isotopes between two phases or molecular species. The equilibrium constants for these reactions may be directly related to isotopic fractionation factors, providing a link between fractionation effects and thermodynamic theory.

The methods of classical thermodynamics may be used to estimate isotopic exchange equilibria in a few favorable cases. The most useful calculations of this type involve the exchange of hydrogen or lithium isotopes, but even these calculations carry substantial uncertainty. For exchange reactions that involve the isotopes of heavier elements, the conventional measurements of the thermodynamic state functions of the relevant substances lack the level of precision needed for the classical calculations to be accurate.

As an alternative, isotopic exchange equilibria can be studied with the methods of statistical thermodynamics. Spectroscopic data, combined with suitable molecular models, can be used to completely characterize the energy states of a substance. This information provides a means to completely predict the thermodynamic properties of the substance, including its fractionation behavior. Very accurate calculations of the isotopic equilibria among gaseous molecules have been made in this manner, but liquids and solids are more difficult to treat, and in most cases their fractionation behavior is best established by direct laboratory measurements. Nevertheless, statistical mechanics provides the theoretical basis for the phenomenon of isotopic fractionation, and quantitatively predicts its dependence on temperature.

A special consequence of statistical theory, known as the "rule of the mean," provides compact relationships for isotopic disproportionation reactions and for the relationships among the vapor pressures of compounds and their isotopomers. The rule may be extended to relate the various isotopic fractionations exhibited by elements that have three or more isotopes.

Isotopic fractionation effects have been widely applied in studies of the origin of materials and of ancient environmental conditions. The best known of these applications is the carbonate paleotemperature method, developed by H. C. Urey, which has revealed considerable information about Earth's ancient climates, and particularly about the conditions and cause of the last ice age.

2.9 Problems

1. The ΔG_{rxn}^0 value for equation 2.6 is equal to -45 cal/mol. Noting that α equals K for this exchange reaction, calculate the value of α to the proper number of significant figures.

2. Assuming ideal mixing in both the vapor and liquid phases, use the information in Table 2.1 to predict the vapor pressure at $25°C$ of an isotopically enriched water sample containing 10 wt. % HDO.

3. Deduce the symmetry numbers for the following molecules: $^{15}N^{14}N$, $^{16}O^{16}O$, $C^{16}O_2$, N_2O, NH_3, C_2H_4, and SF_6.

4. What are the symmetry numbers for methane, CH_4, and the deuteromethanes CDH_3, CH_2D_2, CHD_3, and CD_4? What happens if ^{13}C is present instead of ^{12}C?

5. Calculate the relative abundances of methane and its deuterated isotopomers, using equation 2.8b or 2.8c. Then, for any pair of these methanes, confirm their calculated abundances and confirm equation 2.9. Does CH_2D_2 provide an exception, and, if so, why?

6. How many distinguishable, stable isotopic species exist for the carbonate ion (CO_3^{2-}) and the sulfate ion (SO_4^{2-})?

7. Use the information in Table 2.1 to calculate the equilibrium constant at $25°C$ for the reaction, $H_2 + D_2 = 2HD$. Compare your result with the theoretical value for K_∞ given in equation 2.21, and with Urey's calculated value given in Table 2.4.

8. Use the definition of Q given in equation 2.15 to show that the relationship for the average energy given by equation 2.16 is correct.

9. Use the definition of the Helmholtz free energy, $A = E - TS$, together with the classical relationship $(\partial A/\partial T)_v = -S$, to prove the validity of equation 2.18.

10. According to Table 2.2, the wavenumber $(\omega = 1/\lambda)$ of $^{16}O_2$ is 1580.193 cm^{-1}. Assuming that the molecules behave as simple harmonic oscillators, use the known masses of the oxygen isotopes to predict the wavenumbers for the diatomic species $^{16}O^{17}O$, $^{16}O^{18}O$, and $^{18}O^{18}O$.

11. The SHO model provides a simple relationship (equation 2.30) between the wavenumbers and the reduced masses for pairs of isotopomers. Use the data in Table 2.2 to deduce an approximate relationship between the reduced masses and ratios of $\omega_e \chi_e$, and a relationship between the reduced masses and ratios of the B_e, for pairs of isotopomers.

12. Use the data in Table 2.2 to determine the depth D_e of the energy well for (a) H_2; (b) HD; and (c) D_2 gases. Express your results in cm^{-1}, and also in kcal/mol. Explain your answer.

13. Determine the force constant k for a Morse oscillator by taking the derivative $\partial^2 V / \partial r^2$ of equation 2.36a and evaluating the result at $r = r_e$. Then, introduce this result into equation 2.29d to estimate the value for β in terms of spectroscopic constants. Compare your result to equation 2.36b, given by Huber and Herzberg (1979). Last, using your equation for the force constant, determine the equation for the parabola that approximates the energy well for hydrogen gas.

14. According to the experiments of Clayton et al. (1972), the oxygen isotopic fractionation factor between quartz and water is represented by

$$1000 \ln \alpha_{qtz-w} = 3.38 \times 10^6 / T^2 - 2.90 \qquad \text{range } 200-500°C$$

Calculate α_{qtz-w} for temperatures of 0°C, 25°C, 100°C, and 300°C, recognizing that the first three cases require extrapolation of this empirical equation beyond its range of determination. Then use your results to estimate the $\delta^{18}O$ value of a chert sample that formed at the bottom of the ocean ($\delta^{18}O_w = 0$). What would be the $\delta^{18}O$ value of siliceous sinter formed in a geyser at Yellowstone National Park that erupts water with $\delta^{18}O_w = -16$? Last, calculate the $\delta^{18}O$ value of a quartz vein from the Mother Lode of California, formed at 300°C in hydrothermal fluid with $\delta^{18}O_w = +11$.

15. Use equation 2.61, together with the data in Table 2.2, to estimate the equilibrium constant for the disproportionation reaction

$$H_2 + D_2 = 2HD$$

Compare your result with the theoretical value for K_∞ given by equation 2.21, with the classical thermodynamic result calculated in problem 7, and with Urey's calculated value given in Table 2.4.

16. Use the data in Table 2.2 to determine the equilibrium constant K at 298, 500, and 1000 K for the exchange reaction

$$H_2 + D^{35}Cl = HD + H^{35}Cl$$

Then, calculate the hydrogen isotope fractionation factor α between hydrogen chloride and diatomic hydrogen.

17. At 300 K, the vapor pressure of NH_3 is 10.62 bar, while that of ND_3 is approximately 10.3 bar. In terms of these endmembers, predict the vapor pressures of NHD_2 and NH_2D at 300 K.

18. The fractionation factor for $^{34}S/^{32}S$ partitioning between two sulfides is 1.0160. Use equation 2.74 to predict the fractionations that would be observed for the $^{33}S/^{32}S$, $^{36}S/^{32}S$, and $^{36}S/^{33}S$ systems.

19. Using simple notions about chemical bonds, predict the order of ^{13}C and D enrichment among equilibrated methane, ethane, propane, and butane. Would this order be the same, or the opposite, if the hydrocarbons were kinetically cleaved from complex organic materials undergoing thermal maturation or cracking? In which of these compounds might intramolecular fractionations occur?

20. Use the calcite–water fractionation factor given in Table 2.3 to estimate the change in temperature that would be required to produce the isotopic variations shown in the deep-sea core shown in Figure 2.9, assuming that the oxygen isotope ratio of the water was constant over time. Use equation 2.75 to check your result.

21. Use the ΔG^0_{rxn} value of 2053 cal/mol for the vaporization reaction $H_2O_{liq} = H_2O_{vapor}$, together with the oxygen isotopic fractionation factor α of 1.0094 between water and water vapor at 25°C, to estimate the ΔG^0_{rxn} value for the vaporization reaction $H_2{}^{18}O_{liq} = H_2{}^{18}O_{vapor}$.

References

Bigeleisen, J. (1955) Statistical mechanics of isotope systems with small quantum corrections. I. General considerations and the rule of the geometric mean. *J. Chem. Phys.*, **23**, 2264–2267.

Bigeleisen, J. (1958) The significance of the product and sum rules to isotope fractionation processes. In *Proceedings of the International Symposium on Isotope Separation*, Kistemaker, J., Bigeleisen, J., and Nier, A. O. C., eds. Amsterdam, pp. 121–157.

Bigeleisen, J. (1965) Chemistry of isotopes. *Science*, **147**, 463–471.

Bigeleisen, J. and Mayer, M.G. (1947) Calculation of equilibrium constants for isotopic exchange reactions. *J. Chem. Phys.*, **15**, 261–267.

Bottinga, Y. (1968) Calculation of fractionation factors for carbon and oxygen isotopic exchange in the system calcite–carbon dioxide–water. *J. Phys. Chem.*, **72**, 800–808.

Bottinga, Y. and Javoy, M. (1973) Comments on oxygen isotope geothermometry. *Earth Planet. Sci. Lett.*, **20**, 250–265.

Bottinga, Y. and Javoy, M. (1975) Oxygen isotope partitioning among the minerals in igneous and metamorphic rocks. *Rev. Geophys. Space Phys.*, **13**, 401–418.

Chiba, H., Chacko, T., Clayton, R.N., and Goldsmith, J.R. (1989) Oxygen isotope fractionation involving diopside, forsterite, magnetite, and calcite: application to geochemistry. *Geochim. Cosmochim. Acta*, **53**, 2985–2995.

Clayton, R.N., O'Neil, J.R., and Mayeda, T.K. (1972) Oxygen isotopic exchange between quartz and water. *J. Geophys. Res.*, **77**, 3057–3067.

Criss, R.E. (1991) Temperature dependence of isotopic fractionation factors. *Geochem. Soc. Spec. Pub.*, **3**, 11–16.

Criss, R.E. (1995) Stable isotope distribution: variations from temperature, organic and water–rock interactions. In *Global Earth Physics: A Handbook of Physical Constants*, Ahrens, T. J., ed. AGU Reference Shelf 1, Washington, D.C., pp. 292–307.

Emiliani, C. (1955) Pleistocene paleotemperatures. *J. Geol.*, **63**, 538–578.

Emiliani, C. (1978) The cause of the ice ages. *Earth Planet. Sci. Lett.*, **37**, 349–352.

Epstein, S. and Mayeda, T. (1953) Variation of O^{18} content of waters from natural sources. *Geochim. Cosmochim. Acta*, **4**, 213–224.

Epstein, S., Buchsbaum, R., Lowenstam, H.A., and Urey, H.C. (1953) Revised carbonate–water isotopic temperature scale. *Bull. Geol. Soc. America*, **64**, 1315–1326.

Ferronsky, V.I. and Polyakov, V.A. (1982) *Environmental Isotopes in the Hydrosphere*. John Wiley & Sons, New York.

Friedman, I. and O'Neil, J.R. (1977) Compilation of stable isotope fractionation factors of geochemical interest. In *Data of Geochemistry*, Fleischer, M., ed. U.S. Geological Survey Professional Paper 440 KK.

Friedman, L. and Shiner, V.J. (1966) Experimental determination of the disproportionation of hydrogen isotopes in water. *J. Chem. Phys.*, **44**, 4639–4640.

Gregory, R.T. and Taylor, H.P., Jr. (1981) An oxygen isotope profile in a section of Cretaceous oceanic crust, Samail ophiolite, Oman: evidence for $\delta^{18}O$-buffering of the oceans by deep (> 5 km) seawater-hydrothermal circulation at mid-ocean ridges. *J. Geophys. Res.*, **86**, 2737–2755.

Herzberg, G. (1950) *Molecular Spectra and Molecular Structure. I. Spectra of Diatomic Molecules*. Van Nostrand, New York.

Hill, T.L. (1960) *Introduction to Statistical Thermodynamics*. Addison-Wesley, London.

Huber, K.P. and Herzberg, G. (1979) *Molecular Spectra and Molecular Structure. IV. Constants of Diatomic Molecules*. Van Nostrand Reinhold, New York.

Jones, T.F. (1958) Vapor pressures of some isotopic substances. In *Proceedings of the International Symposium on Isotope Separation*, Kistemaker, J., Bigeleisen, J., and Nier, A.O.C., eds. (Amsterdam, 1957), pp. 74–102.

Kieffer, S.W. (1982) Thermodynamics and lattice vibrations of minerals: 5. Applications to phase equilibria, isotopic fractionation, and high pressure thermodynamic properties: *Rev. Geophys. Space Phys.*, **20**, 827–849.

Kajiwara, Y. and Krouse, H.R. (1971) Sulfur isotope partitioning in metallic sulfide systems. *Can. J. Earth Sci.*, **8**, 1397–1408.

Kirshenbaum, I. (1951) In *Physical Properties and Analysis of Heavy Water*, Urey, H.C. and Murphy, G.M., eds. McGraw-Hill Book Co., New York.

Lide, D.R. (1991) *Handbook of Chemistry and Physics*, 71st ed. CRC Press, Boston, Massachusetts.

McCrea, J.M. (1950) On the isotopic chemistry of carbonates and a paleotemperature scale. *J. Chem. Phys.*, **18**, 849–857.

Majoube, M. (1971) Fractionnement en oxygene 18 et en deuterium entre l'eau et sa vapeur. *J. Chim. Phys.*, **68**, 1425–1436.

Melchiorre, E.B., Criss, R.E., and Rose, T.P. (1999) Oxygen and carbon isotope study of natural and synthetic malachite. *Econ. Geol.*, **94** (in press).

Merlivat, L., Botter, R., and Nief, G. (1963) Fractionnement isotopique au cours de la distillation de l'eau. *J. Chim. Phys.*, **60**, 56–59.

Mills, G.A. and Urey, H.C. (1940) The kinetics of isotopic exchange between carbon dioxide, bicarbonate ion, carbonate ion and water. *J. Am. Chem. Soc.*, **62**, 1019–1026.

Mook, W.G., Bommerson, J.C., and Staverman, W.H. (1974) Carbon isotope fractionation between dissolved bicarbonate and gaseous carbon dioxide. *Earth Planet. Sci. Lett.*, **22**, 169–176.

Morse, P.M. (1929) Diatomic molecules according to the wave mechanics. II. Vibrational levels. *Phys. Rev.*, **34**, 57–64.

Muehlenbachs, K. and Clayton, R.N. (1976) Oxygen isotope composition of the oceanic crust and its bearing on seawater. *J. Geophys. Res.*, **81**, 4365–4369.

O'Neil, J.R. (1986) Theoretical and experimental aspects of isotopic fractionation. *Rev. Mineral.*, **16**, 1–40.

O'Neil, J.R., Clayton, R.N., and Mayeda, T.K. (1969) Oxygen isotope fractionation in divalent metal carbonates. *J. Chim. Phys.*, **51**, 5547–5558.

Richet, P., Bottinga, Y., and Javoy, M. (1977) A review of hydrogen, carbon, nitrogen, oxygen, sulphur, and chlorine stable isotope fractionation among gaseous molecules. *Ann. Rev. Earth Planet. Sci.*, 65–110.

Stern, M.J., Spindel, W., and Monse, E.U. (1968) Temperature dependence of isotope effects. *J. Chem. Phys.*, **48**, 2908–2919.

Urey, H.C. (1947) The thermodynamic properties of isotopic substances. *J. Chem. Soc.* (London), 562–581.

Wachter, E.A. and Hayes, J.M. (1985) Exchange of oxygen isotopes in carbon dioxide–phosphoric acid systems. *Chem. Geol.*, **52**, 365–374.

3

Isotope Hydrology

3.1 Variations of D and ^{18}O in the Hydrosphere

No substance exemplifies the principles of isotope distribution better than water. Water is practically ubiquitous at the Earth's surface, where it undergoes phase transitions, interacts with minerals and the atmosphere, and participates in complex metabolic processes essential to life. The isotopes of hydrogen and oxygen undergo large fractionations during these processes, providing a multiple isotopic tracer record of diverse phenomena.

In the hydrologic cycle, hydrogen and oxygen isotope ratios provide conservative tracers, uniquely intrinsic to the water molecule, that elucidate the origin, phase transitions, and transport of H_2O. In particular, the isotope data associated with these processes are amenable to theoretical modeling using the laws of physical chemistry.

The characteristics of the principal reservoirs of natural waters on Earth are provided in the following sections. The distinct characters of these different reservoirs are very clearly shown on graphs where the δD values are plotted against those of $\delta^{18}O$ (Figure 3.1).

3.1.1 Ocean Water

The oceans constitute 97.25% of the hydrosphere, cover 70% of the Earth's surface to a mean depth of 3.8 km, and have an enormous total volume of 1.37×10^9 km^3 (Table 3.1). This large reservoir has strikingly uniform isotopic concentrations, with almost all samples having $\delta^{18}O = 0 \pm 1$ and $\delta D = 0 \pm 5$ per mil relative

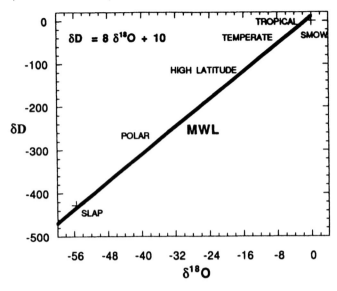

Figure 3.1 The δD versus $\delta^{18}O$ values of meteoric waters generally lie very close to the meteoric water line (MWL). The waters generally become increasingly depleted in D and ^{18}O as one progresses from the tropics to temperate zones to high-latitude regions. The SMOW and SLAP (standard light antarctic precipitation) isotopic standards are indicated by the crosses; the latter has a value of (−55.5, −428) relative to SMOW.

to SMOW (Craig and Gordon, 1965). Values outside these ranges are almost invariably confined to surface waters that have salinities that differ from the normal value of 3.5 wt. %. These varations are generally attributable to evaporation, formation of sea ice, or addition of meteoric precipitation that may occur by direct rainfall, by river inflow, or by melting of icebergs. The latter effect was clearly documented by Epstein and Mayeda (1953) in the surface waters of the

Table 3.1 Relative Volumes and Typical δD and $\delta^{18}O$ Values of Hydrospheric Reservoirs

Reservoir	Volume (%)	δD(‰)	$\delta^{18}O$ (‰)
Ocean	97.2	0 ± 5	0 ± 1
Ice caps and glaciers	2.15	-230 ± 120	-30 ± 15
Groundwater	0.62		
Vadose water		-40 ± 70	-5 ± 15
Dilute groundwater		-50 ± 60	-8 ± 7
Brines		-75 ± 50	0 ± 4
Surface waters	0.017		
Freshwater lakes		-50 ± 60	-8 ± 7
Saline lakes and inland seas		-40 ± 60	-2 ± 5
River and stream channels		-50 ± 60	-8 ± 7
Atmospheric water	0.001	-150 ± 80	-20 ± 10

North Atlantic, where the isotopic variations were strongly correlated with variations in salinity.

In detail, the deep waters of different ocean basins have distinct values of $\delta^{18}O$ and salinity. Thus, the $\delta^{18}O$ values of deep waters from the North Atlantic (ca. +0.05‰), Pacific (−0.15‰), and Antarctic (−0.40‰) oceans are distinct, and careful measurements can be used to infer details of oceanic circulation patterns (Craig and Gordon, 1965).

While some workers have argued that a large secular change in the $\delta^{18}O$ values of ocean water has occurred over geologic time, it is much more likely that a value near zero is maintained by fluid–rock interactions at spreading centers (Muehlenbachs and Clayton, 1976). In effect, the $\delta^{18}O$ value of the hydrosphere is buffered by the Earth's lithosphere, on a timescale of ~30 Ma that is required to circulate the entire ocean through the mid-ocean hydrothermal systems (e.g., Gregory and Taylor, 1981). In contrast, preferential loss of protium from the upper atmosphere to space may have resulted in a significant change in the δD value of the hydrosphere over geologic time.

On a shorter timescale, fluctuations of approximately ±1 and ±8 per mil in the $\delta^{18}O$ and δD values of ocean water, respectively, occurred during the Quaternary period. These fluctuations clearly relate to cycles of growth and melting of continental ice sheets (e.g., Savin and Yeh, 1981), and occurred on a timescale that was much too fast to be offset by lithospheric buffering. It is interesting to predict what would happen if the great Antarctic and Greenland ice caps were completely melted today. Because of their enormous volume, sea level would rise by nearly 100 m. Assuming that the mean $\delta^{18}O$ value of this ice is about −35 ± 5‰, the material balance relation (equation 1.26a) can be used to calculate that the $\delta^{18}O$ value of the ocean would change from 0 to about −0.7 ± 0.1‰.

3.1.2 Meteoric Waters

Meteoric waters originate as atmospheric precipitation through the operation of the hydrologic cycle. This water is very familiar to us as rain and snow, and in its derivative forms of streamflow, lake water, soil water, glacier ice, and shallow groundwater.

The physical processes responsible for the production, transport, and condensation of atmospheric water vapor cause large variations in the isotopic ratios of meteoric water. The total range of $\delta^{18}O$ values in natural precipitation is about +4 to −62‰, while that of the δD values is about +40 to −500‰. These values vary in a highly characteristic manner in practically all meteoric waters of the world. Specifically, modern meteoric waters conform closely to an empirical relationship known as the "meteoric water line," or MWL (Figure 3.1; Craig, 1961):

$$\delta D = 8\,\delta^{18}O + 10$$

(3.1)

Note that SMOW does not lie on the MWL; the +10 intercept is defined as the "deuterium excess" (Figure 3.1). To first order, this intercept is produced by kinetic effects that accompany the evaporation of seawater (see chapter 4; Craig and Gordon, 1965). In detail, the precise value of the deuterium excess varies geographically, and "local meteoric water lines" that have different intercepts and slightly different slopes have been proposed for different regions. For example, a meteoric water line that has a deuterium excess of $+18$ or more better characterizes the precipitation of the eastern Mediterranean region than does equation 3.1 (Gat and Carmi, 1970).

3.1.2.1 Meteoric Precipitation   The systematic variations in the δD and $\delta^{18}O$ values of meteoric precipitation elucidate the atmospheric conditions of their origin, and serve as invaluable guides to the source regions of derivative waters such as springs and groundwaters. The isotopic values of meteoric precipitation are principally correlated with temperature, but they also vary with altitude, latitude, proximity to the ocean, and other factors (Dansgaard, 1964). Atmospheric air masses tend to become increasingly depleted in the heavy isotopes D and ^{18}O, during their progressive condensation and removal of precipitation. As a result, the $\delta^{18}O$ values of rain in the tropics (mostly $+1$ to -3) is similar to ocean water, but the values become systematically lower (mostly -3 to -12) in temperate zones, are even lower in continental interiors at high latitudes, and ultimately attain extremely low values (to -62) in winter ice at the South Pole (Figure 3.1).

The δD and $\delta^{18}O$ values of meteoric waters are strongly correlated with air temperature. Dansgaard (1964) provides the following empirical equation that relates the average $\delta^{18}O$ values of precipitation to the mean annual surface air temperature of a given locality, where T_{avg} is in degrees centigrade:

$$\delta^{18}O \cong 0.695 T_{avg} - 13.6 \qquad (3.2)$$

This relationship is only a rough approximation, and it largely represents a fit to data from European and North Atlantic stations. Data for North America suggest a similar temperature coefficient, but, at a given temperature, the $\delta^{18}O$ values would typically be 1–4‰ lower than would be predicted by equation 3.2. Thus, the temperature coefficient is approximately

Empirical temperature effect: $\partial \delta^{18}O / \partial T \cong 0.7‰/deg$ $\qquad (3.3a)$

Increasing elevation also causes meteoric precipitation to become progressively depleted in heavy isotopes. Several authors have measured such effects: along the western flank of the Sierra Nevada, for example, Rose et al. (1996) found that the $\delta^{18}O$ values systematically decrease by 2.3‰ for every vertical kilometer of elevation gain. Thus, the empirical "altitude effect" is approximately

Empirical altitude effect: $\partial \delta^{18}O / \partial z \cong 2.0 \pm 1‰/km$ $\qquad (3.3b)$

The distance from the ocean, or the "continentality" of a site, seems to cause an additional but highly variable effect on the isotopic concentrations. Across Eurasia, the $\delta^{18}O$ values of meteoric precipitation decrease by about 0.35‰ per 100 km of lateral distance (Ferronsky and Polyakov, 1982). On the other hand, across the African Sahel the gradient is approximately 0.084‰ per 100 km (Joseph et al., 1992). Even more variable are gradients that are as great as

7.5‰ per 100 km across northern California, yet negligible across Nevada (Ingraham and Taylor, 1991), although the former region incorporates effects due to a significant change in altitude, while the second represents the interesting case of a hydrologically closed basin. In any case, longitudinal effects may principally reflect the amount of rainout that occurs, on average, along the predominant storm tracks. Neglecting special cases, and eliminating altitude effects, the observed longitudinal gradients are mostly

$$\text{Empirical longitude effect:} \qquad \partial \delta^{18}O/\partial x \cong 0.002 \pm 0.002‰/km \qquad (3.3c)$$

Last, the so-called "latitude effect" could be a combination of rainout and decreasing temperatures. The typical magnitude is

$$\text{Empirical latitutde effect:} \qquad \partial \delta^{18}O/\partial x \cong 0.002 \pm 0.001‰/km \qquad (3.3d)$$

Latitudinal gradients as high as 2.2‰ per 100 km occur poleward in Antarctica, although these effects incorporate changes in altitude and continentality.

3.1.2.2 Glaciers and Ice Caps   Ice constitutes 2.15% of the hydrosphere and represents 70% of the freshwater available on Earth. While the $\delta^{18}O$ and δD values of ice vary with temperature, altitude, latitude, and so on, in a manner similar to ordinary meteoric waters, glaciers and ice caps are of particular interest because they contain the most isotopically depleted precipitation on Earth. Moreover, in areas of snow and ice accumulation, the isotopic ratios of ice provide valuable information on seasonal temperature variations, the physical maturation of snow to ice, and glacier flow.

Snow packs exhibit pronounced isotopic variations with depth that record seasonal temperature fluctuations, with the most depleted samples corresponding to winter (Figure 3.2). Such variations help to establish stratigraphic relationships in snow packs, that may, in turn, be used to define climatic variations. Over time, as accumulation continues and the snow pack metamorphoses into ice, earlier annual layers become attenuated, while diffusion and recrystallization processes result in a progressive reduction of the amplitude of the isotopic variations (Johnsen et al., 1972). Even after small-scale isotopic homogenization of the ice has occurred, the isotopic ratios of the ice can help to define the flow mechanics of flowing glaciers, or the long-term climatic variations of individual sites (see later).

3.1.2.3 Groundwaters   Dilute, cool, shallow groundwaters are extremely important to man. They provide nearly half of the domestic water supply in Europe and the United States, and are even more important in arid regions. The origin of these waters and the processes that affect them are therefore of great practical importance, and many of these aspects are elucidated by stable isotope data.

Most shallow groundwaters lie along or close to the MWL. Groundwater systems generally do a remarkable job of integrating the local precipitation from which they originate, and so provide an excellent weighted average of the δD and $\delta^{18}O$ values of that precipitation. There are a few exceptions to this useful generalization. First, in some regions, particularly semiarid ones, only the largest, longest-lived storm events recharge the groundwater systems, and the δD and

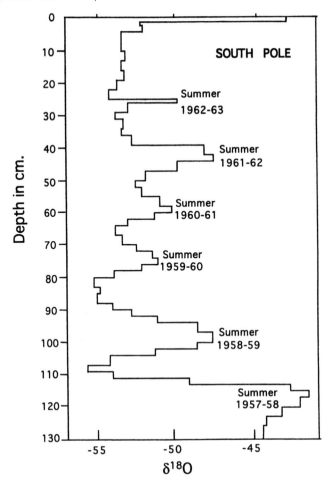

Figure 3.2 Seasonal variations of the $\delta^{18}O$ values in the snow pack at the South Pole, after Epstein et al. (1965). The $\delta^{18}O$ values are very low, especially in winter, and the average accumulation rate is only 7.1 cm per year of water equivalent.

$\delta^{18}O$ values of those storms may differ from the overall average. Second, evaporation processes in the vadose zone or in alpine snowpacks can cause isotopic fractionation of the water, and in detail can produce complicated variations that may include a decrease in the deuterium excess. Third, some groundwaters, such as those in artesian basins, have clearly migrated tens or hundreds of kilometers from higher elevation points of recharge, and accordingly they can have lower δD and $\delta^{18}O$ values than the local precipitation at their sties of collection. Last, in some cases, groundwaters were recharged during ancient pluvial periods, when the meteoric precipitation had different average values and a lower deuterium excess than today.

In large clastic basins, the δD and $\delta^{18}O$ values of the groundwaters commonly decrease with depth. This effect may represent the longer, slower flowpaths of progressively deeper waters, which therefore have a greater age and were recharged at progressively higher elevations, consistent with hydrologic models. In addition, the effects of man may profoundly affect the recharge patterns, quality, and isotope ratios of unconfined groundwater systems, particularly in urbanized or irrigated areas.

3.1.2.4 Surface Waters   The waters of freshwater lakes and flowing streams, though highly familiar, constitute only a minute fraction (< 0.02%) of the hydrosphere. The δD and δ^{18}O values of these waters predictably indicate their meteoric parentage, because these waters volumetrically integrate the parent meteoric waters that fall within their watersheds. Far more interestingly, the isotopic data illustrate that complex processes accompany the formation of these waters.

One important process is surface evaporation, which affects all surface waters, but especially shallow lakes. Of course, such effects are most pronounced in windy, hot, or arid areas. The kinetic effects associated with evaporation typically cause the waters to plot significantly to the right of the MWL, as will be discussed in detail in chapter 4.

An unexpected effect is the extent to which surface waters are linked with shallow groundwater systems. In gaining streams and in many discharge lakes, significant fractions of the flow are contributed by groundwater. Most of the streams and lakes in temperate areas are of these respective types, but, in arid and semiarid regions, losing streams and recharge lakes supply water *to* the groundwater systems. Isotopic studies elucidate these processes with unparalleled clarity, because the isotopic ratios of individual storms are unique, and, in general, deviate from the ratios of the average precipitation, groundwater, and streamflow. Several studies have now proven that 50% or more of the flowing water observed in most streams and rivers represents "pre-event" water, even during the spectacular rises in streamflow that accompany major storm events. One of the principal effects of rainfall, then, is to displace groundwater into stream channels.

3.2 Variations of D and ^{18}O in Waters from Deep Geologic Environments

Important aqueous fluids that have diverse origins also occur more deeply within the Earth. Because these fluids have resided for long periods within rocks at elevated temperatures, they generally have higher salinities than ordinary meteoric waters. Examples of these fluids include the formation waters encountered in oil fields, the waters and steams of geothermal systems and volcanoes, and fluids principally known only from microscopic inclusions in crystalline rocks and ore deposits.

Typically, the δD and δ^{18}O values of these deep waters lie well to the right of the MWL, a result of their modest to extensive exchange with isotopes in the host rocks (Figure 3.3). In some important cases, the δD and δ^{18}O values of these deep waters allow them to be recognized as evolved meteoric waters or trapped seawaters. Some of these demonstrations have literally revolutionized understanding

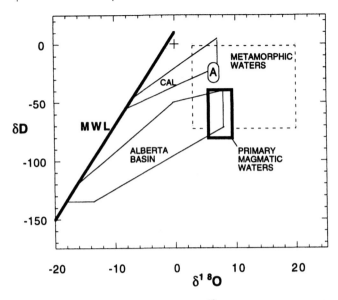

Figure 3.3 Graph showing fields of δD versus $\delta^{18}O$ values calculated for magmatic and metamorphic waters, and measured for representative oil-field brines from California and the Alberta basin. The small field labeled "A" represents "Andesite water" identified at certain active volcanoes. The MWL (heavy line) and SMOW (cross) are shown for reference. Modified after Sheppard (1986).

of ore deposits and geothermal systems. However, for the most "highly evolved" waters, representing waters that have completely equilibrated with rocks at high temperatures, the δD and $\delta^{18}O$ values are so profoundly modified that they can no longer be used to deduce the fluid origin. Only the conditions of the fluid–rock interactions that control the δD and $\delta^{18}O$ values of re-equilibrated waters can be deduced in such cases.

3.2.1 Formation Waters

Formation waters typically occur in the pore volumes of rocks in sedimentary basins, where temperatures are mostly 30–150°C. These saline waters are commonly encountered by deep drilling in oil fields, but in regions undergoing tectonic compression they can appear as saline surface springs (Davisson et al., 1994; see below). Formation waters typically have a complex history, and they occupy a continuum that extends from dilute, shallow meteoric groundwaters to highly evolved metamorphic waters. Many formation fluids originated as seawater trapped in the pore volumes of accumulating sediments, but commonly they have undergone extensive interactions with rocks or been diluted with shallow meteoric groundwaters.

Formation waters contain small to very substantial concentrations of dissolved constituents. While many of these dissolved minerals have a connate (seawater) origin, even larger amounts may result from fluid–rock interactions, sometimes including interactions with evaporites. Some oil-field waters range into hot (> 150°C), highly concentrated (> 10 wt. %) brines. The dominant anion in formation waters is normally Cl⁻, while the major cations are typically Ca^{2+} and Na^+. The relative proportion of Ca^{2+} typically increases with increasing salinity and temperature. Davisson and Criss (1996) have demonstrated, for formation fluids collected throughout the world, that the Ca^{2+}, Na^+, and Cl⁻ concentrations share a quantitative balance as a result of Ca^{2+} and Na^+ exchange of rocks with a parent fluid that originally had ion ratios similar to seawater.

Fluid–rock interactions profoundly affect the δD and $\delta^{18}O$ values of formation waters. In cases where fluids at low temperature are in contact with highly reactive materials, the $\delta^{18}O$ values may decrease by several per mil, and the fluid may plot slightly to the *left* of the MWL. Such effects have been observed in shallow marine sediments (Lawrence and Gieskes, 1981) and rarely in volcanic environments (Kelley et al., 1986). Far more commonly, the δD versus $\delta^{18}O$ values plot along positive-sloped trends that appear to "originate" from a point on the MWL that is appropriate for local precipitation, and extend toward a cluster of higher δD and $\delta^{18}O$ values that represent the hottest, most saline samples (Figure 3.3). Clayton et al. (1966) argued that these fluids originated from meteoric waters, and were modified by fluid–rock exchange during their protracted residence in the subsurface. Several subsequent studies have demonstrated that these fluids more commonly represent trapped, evolved seawater that has undergone variable amounts of meteoric dilution, a process that produces trends that have a similar appearance but evolve in the reverse direction. In many cases, particularly where the fluid is overpressured, the fluids are clearly modified connate waters (Kharaka and Carothers, 1986).

3.2.2 Hydrothermal Fluids

Hydrothermal fluids are hot, aqueous fluids of any origin. The most spectacular examples occur in geothermal areas, where their circulation and discharge is primarily driven by buoyancy forces. While the surface temperatures of hot systems are necessarily controlled by the boiling point, the temperatures of the subsurface reservoirs commonly approach the critical point of water. The two most important types of hot fluid observed in geothermal areas are chloride-type waters and acid-sulfate waters.

Chloride-type geothermal waters are the volumetrically dominant type of hydrothermal fluid. These waters have near-neutral pH, and have substantial salinities, typically thousands of grams per kilogram, that are dominated by Na^+, K^+, Ca^{2+}, and Cl⁻. They also are close to saturation with silica, and emanate from free-flowing hot springs or, in spectacular cases, from geysers surrounded by silica terraces (White et al., 1971).

In an important study, Craig et al. (1956) unequivocally proved that chloride-type hydrothermal fluids are dominantly derived from the local meteoric waters in each area of occurrence. The key observation is that the hot fluids typically share the same D/H ratio as local meteoric waters, which vary from one geothermal system to another (Figure 3.4). However, the fluids are generally higher in ^{18}O than the local meteoric waters. The latter effect, known as the "^{18}O-shift," reflects

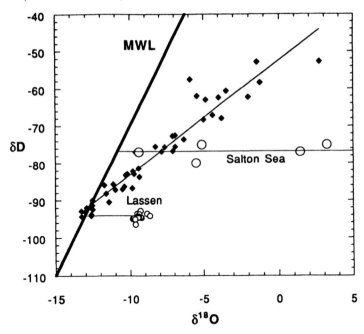

Figure 3.4 Graph of δD versus $\delta^{18}O$ values of some hydrothermal fluids from the Lassen (Janik et al., 1983) and the Salton Sea (Craig, 1966) geothermal regions. Near-neutral, chloride-type (open circles) and acid-sulfate (filled diamonds) geothermal fluids from Lassen Volcanic National Park define lines that clearly originate on the MWL at a point identical to the local meteoric waters at that locality. However, the chloride waters at Lassen and at the Salton Sea exhibit significant, zero-slope "^{18}O-shifts" due to water–rock interactions. In contrast, the trend for the acid-sulfate waters at Lassen (slope ~3.07) is characteristic of evaporation at near-boiling temperatures.

interaction and exchange with rocks, which are much richer in ^{18}O (Figure 3.4). Thus, after centuries of argument about their nature, isotopic studies have provided definitive evidence that these fluids are not primary emanations from the deep Earth, but rather are ordinary surface waters that have deeply penetrated the rocks and become heated.

Acid-sulfate waters are chemically aggressive fluids in which H^+ is the dominant cation and SO_4^{2-} is the dominant anion. These waters accordingly have very low pH (commonly < 2.5), high SO_4^{2-} but low Cl^- contents, and are typically associated with steaming ground, mud pots, and fumaroles (White et al., 1971). The δD and $\delta^{18}O$ trend lines characteristic of these waters also originate on the MWL at points appropriate for the local meteoric precipitation. These waters therefore also appear to have originated from local meteoric waters, but the low (2–3) positive slopes of the isotopic trend lines indicate their modification by evaporative (kinetic) effects at near-boiling temperatures (Figure 3.4). Acid-sulfate waters commonly occur proximal to chloride-type waters, and may represent condensed steam derived from deep chloride-type reservoirs (White et al., 1971).

The hot (to 350°C) chloride-rich fluids that discharge from mid-ocean ridge spreading centers are clearly derived from seawater, but have undergone changes in chemistry, including complete Mg loss, along with increases of about 2‰ in the δD and $\delta^{18}O$ values (e.g., Bowers and Taylor, 1985).

3.2.3 Magmatic Waters

Magmatic water is water of any origin that has equilibrated with magma. Such waters are important in the genesis of igneous rocks, and essential to the development of certain types of ore deposits, such as porphyry Cu and Mo deposits, Sn and W greisens, and pegmatites (e.g., Sheppard et al., 1971; Kamilli and Criss, 1996). The δD (−40 to −80) and $\delta^{18}O$ (+5.5 to +9.5) values of this water may be estimated from fluid inclusion data and from calculations that assume equilibrium with crystalline rocks at igneous temperatures (Sheppard, 1986). Any fluid that undergoes extensive interaction with crystalline rocks at elevated temperatures would develop similar characteristics, so such waters may be derived from subducted or deep-circulating surface waters. Taylor (1986) provides a useful discussion of magmatic volatiles.

The term *"primary magmatic water"* is reserved for water that has been transported within a magma from a deep-seated source (Sheppard et al., 1971). Such waters may have been observed in, for example, Satsuma Iwojima volcano in southwest Japan, where > 700°C fumaroles emanate "andesite water" that has a $\delta^{18}O$ value of +6.5, a δD value of −25, and a chlorinity of 12,000 ppm (Hedenquist et al., 1994). In most regions on Earth, any such waters are volumetrically dwarfed by circulating, heated waters derived from surface environments.

3.2.4 Metamorphic Waters

Metamorphic water has been equilibrated with metamorphic rocks at elevated temperatures, and can be viewed as an extreme type of formation water. Large quantities of such waters are only rarely observed to discharge from the Earth's surface, but they transport important quantities of heat and chemicals in deep metamorphic environments, and arguably are essential to the development of certain ore deposits. The δD (0 to −70) and $\delta^{18}O$ (+3 to +20) values of metamorphic waters may be calculated by assuming equilibrium with metamorphic rocks at appropriate temperatures (e.g., Sheppard, 1986).

3.2.5 Juvenile Waters

These hypothetical waters presumably have never undergone interactions with the hydrosphere, and therefore must originate from the deep interior of the Earth. Many geochemical models suggest that the entire hydrosphere originated from the outgassing of such waters from the solid Earth, and that this outgassing continues today while the oceans increase in volume. There are no convincing samples of waters of this type.

3.3 Liquid–Vapor and Ice–Vapor Equilibria

The key process governing the hydrologic cycle is the phase change between water vapor and condensate. The isotopic fractionations associated with this process are primarily responsible for the production of the δD and $\delta^{18}O$ variations in natural waters. This section will demonstrate the straightforward relationships between the liquid–vapor fractionation factors and the vapor pressures of the various isotopomers of water.

3.3.1 Elementary Meteorological Relationships

Numerous handbooks tabulate the variation of the saturated vapor pressure of water with temperature, or provide accurate power series for its computation. Alternatively, approximate values may be calculated from the Clausius-Clapeyron equation (equation 2.5c), using a heat of vaporization centered in the temperature range of interest. Thus, for temperatures of 0–25°C,

$$\ln P_{sat} \cong 21.113 - \frac{5350.3}{T} \tag{3.4a}$$

where T is temperature in kelvins and the vapor pressure of water is given in torr. A similar relationship may be obtained for the vapor pressure above ice, at sub-zero temperatures:

$$\ln P_{sat,ice} \cong 24.0268 - \frac{6147.3}{T} \tag{3.4b}$$

The observed partial pressures, P_{H_2O}, of water vapor in real air masses are commonly lower than the saturated values, P_{sat}. The quantities known as the dew point and the humidity are then of interest. The dew point is the temperature to which the air mass would need to be cooled for its water vapor to begin to condense. At the dew point, the actual vapor pressure of the air is equal to the saturated vapor pressure given by equation 3.4a. The humidity h is defined as the ratio of the actual vapor pressures to the saturated vapor pressure at the actual temperature of the given air mass:

$$\boxed{h = P_{H_2O}/P_{sat}} \tag{3.5}$$

Finally, during the progressive condensation of an air mass, neither the temperature, pressure, nor volume normally will remain constant, and even the total mass of the air will change due to the removal of the condensate. It is therefore useful to introduce a meteorological quantity known as the "mixing ratio," r, which normalizes the water content of an air mass to its content of "dry air" (mostly N_2, O_2, and Ar):

$$\boxed{r = \left(\frac{m_{H_2O}}{m_{dry\,air}}\right)\left(\frac{P_{H_2O}}{P_{total} - P_{H_2O}}\right)} \tag{3.6}$$

In this expression, the first term on the right-hand side represents the ratio of the molecular weights of the water vapor and the dry air, which is 18.0/28.97. This value is conveniently close to the simple fraction, 5/8. The second term on the right-hand side represents the ratio of the partial pressure of water to the partial pressure of the dry air, with the latter representing the difference between the total air pressure and the vapor pressure.

3.3.2 Oxygen Isotope Fractionation

Consider the vapor pressure of $H_2^{18}O$, denoted by $P_{H_2^{18}O}^v$, above a solution composed of $H_2^{16}O$ and $H_2^{18}O$. According to Dalton's law of partial pressures (first equality, below), that partial pressure is equal to the total vapor pressure P_T above the solution, multiplied by the mole fraction of $H_2^{18}O$ in the vapor, $X_{H_2^{18}O}^v$. In addition, according to Raoult's law (second equality), that vapor pressure is also equal to the vapor pressure of pure $H_2^{18}O$, notably the quantity $P_{H_2^{18}O}^0$, multiplied by the mole fraction of $H_2^{18}O$ in the liquid water, $X_{H_2^{18}O}^w$. Insofar as mixtures of the isotopomers of water are almost perfectly ideal, this gives the following virtually exact relationship (S. M. Savin, pers. comm., 1973):

$$P_{H_2^{18}O}^v = X_{H_2^{18}O}^v P_T = X_{H_2^{18}O}^w P_{H_2^{18}O}^0 \qquad (3.7a)$$

Now, consider the vapor pressure $P_{H_2^{16}O}^v$ above the solution. Dalton's law and Raoult's law may similarly be used to relate that vapor pressure to the mole fractions of $H_2^{16}O$ that occur in the vapor phase and in the liquid solutions, notably $X_{H_2^{16}O}^v$ and $X_{H_2^{16}O}^w$. The total vapor pressure, P_T, and the vapor pressure of pure liquid $H_2^{16}O$, $P_{H_2^{16}O}^0$, are also involved in a result similar to equation 3.7a; notably,

$$P_{H_2^{16}O}^v = X_{H_2^{16}O}^v P_T = X_{H_2^{16}O}^w P_{H_2^{16}O}^0 \qquad (3.7b)$$

Now, suppose that equation 3.7a is divided by equation 3.7b. The quotient of the mole fractions of the middle terms is identical to the $^{18}O/^{16}O$ ratio of the vapor, while the P_T terms cancel. Similarly, the quotient of the mole fractions on the right-hand side is identical to the $^{18}O/^{16}O$ ratio of the liquid solution. Thus,

$$\left(\frac{^{18}O}{^{16}O}\right)_v = \left(\frac{^{18}O}{^{16}O}\right)_w \frac{P_{H_2O^{18}}^0}{P_{H_2O^{16}}^0} \qquad (3.7c)$$

Of course, the liquid–vapor fractionation factor, α_{w-v}^{18}, is defined as the quotient R_w/R_v. Thus, rearranging equation 3.7c gives the useful final relationship (S. M. Savin, pers. comm., 1973),

$$\boxed{\alpha_{w-v}^{18} = \frac{P_{H_2O^{16}}^0}{P_{H_2O^{18}}^0}} \qquad (3.8)$$

This important equation demonstrates that the liquid–vapor fractionation factor is equal to the ratio of vapor pressures that would be exhibited above pure liquid

$H_2^{16}O$ and above pure liquid $H_2^{18}O$. At 25°C, this ratio has the value of 1.0094, and it systematically changes as a function of temperature (see Table 3.2).

3.3.3 Hydrogen Isotope Fractionation

For the hydrogen isotopes, equations analogous to equations 3.7a, b, and c for oxygen may be evaluated. The relevant equations become

$$P_{HDO}^v = X_{HDO}^v P_T = X_{HDO}^w P_{HDO}^0 \tag{3.9a}$$

and

$$P_{H_2O}^v = X_{H_2O}^v P_T = X_{H_2O}^w P_{H_2O}^0 \tag{3.9b}$$

where P_{HDO}^0 refers to the vapor pressure that would be exhibited by pure liquid HDO and the other symbols are analogous to those given previously. Now, dividing equation 3.9a by equation 3.9b,

$$\left(\frac{X_{HDO}}{X_{H_2O}}\right)_v = \left(\frac{X_{HDO}}{X_{H_2O}}\right)_w \left(\frac{P_{HDO}^0}{P_{H_2O}^0}\right) \tag{3.9c}$$

Because there are two sites for hydrogen in the water molecule, it is not immediately clear how to relate the ratio of the mole fractions of HDO and H_2O to the D/H ratio, and this is required to complete the analysis. The required relationship can be found by assuming a random distribution of the hydrogen isotopes among the various isotopomers, and using equations 2.8a and b. Letting x represent the abundance of protium (H) in water and y the abundance of D in the water, the relative abundances of the various molecular species become

$$H_2O = x^2 \qquad D_2O = y^2 \qquad HDO + DHO = 2xy \tag{3.10}$$

The D/H ratio of the water may now be computed by adding all the deuterium atoms in all the relevant species, and dividing the result by the total number of protium atoms in all the species. Account must also be made of the fact that HDO has one deuterium atom while D_2O has two; similarly, HDO has one protium atom while H_2O has two. Thus,

$$\left(\frac{D}{H}\right) = \frac{1(HDO) + 2(D_2O)}{1(HDO) + 2(H_2O)} = \frac{2xy + 2y^2}{2xy + 2x^2} = \frac{y}{x} \tag{3.11a}$$

It is clear that the D/H ratio is equal to the quantity y/x, as might have been expected. Further considerations show that the ratio of the mole fractions of the HDO and H_2O molecules is numerically equal to twice the D/H ratio; that is,

$$\frac{X_{HDO}}{X_{H_2O}} = \frac{2xy}{x^2} = 2\left(\frac{y}{x}\right) \tag{3.11b}$$

This last result could also have been obtained using equation 2.9. Thus, equation 3.9c becomes

$$2\left(\frac{D}{H}\right)_v = 2\left(\frac{D}{H}\right)_w \left(\frac{P_{HDO}^0}{P_{H_2O}^0}\right) \tag{3.11c}$$

This simplifies to the final result:

$$\alpha_{w-v}^{D} = \frac{P_{H_2O}^{0}}{P_{HDO}^{0}} \qquad (3.12)$$

where α_{w-v}^{D} refers to the hydrogen isotopic fractionation factor for liquid-water and vapor, representing the quotient the quotient R_w/R_v for hydrogen. At 25°C, the value for α_{w-v}^{D} has the value of 1.0794 (Table 3.2), while the classical thermodynamic calculations in chapter 2 gave a value of 1.079 for the vapor pressure ratio (equations 2.7a, b, and c and following).

The relationships for the D_2O/H_2O pair are more complex, and, in fact, are not as accurate, but everything is in place for their evaluation. Dalton's and Raoult's Laws provide the following two relationships:

$$P_{D_2O}^{v} = X_{D_2O}^{v}P_T = X_{D_2O}^{w}P_{D_2O}^{0} \qquad (3.13a)$$

and

$$P_{H_2O}^{v} = X_{H_2O}^{v}P_T = X_{H_2O}^{w}P_{H_2O}^{0} \qquad (3.13b)$$

Dividing equation 3.13a by equation 3.13b gives

$$\left(\frac{X_{D_2O}}{X_{H_2O}}\right)_v = \left(\frac{X_{D_2O}}{X_{H_2O}}\right)_w \left(\frac{P_{D_2O}^{0}}{P_{H_2O}^{0}}\right) \qquad (3.13c)$$

Now, using the molecular abundances provided in equation 3.10 to evaluate the mole fractions—that is, by substituting x^2 for the mole fraction of H_2O and y^2 for D_2O—gives

$$\left(\frac{y^2}{x^2}\right)_v \cong \left(\frac{y^2}{x^2}\right)_w \left(\frac{P_{D_2O}^{0}}{P_{H_2O}^{0}}\right) \qquad (3.13d)$$

Now, introducing the equivalence of y/x and the D/H ratio established in equation 3.11a gives

$$\left(\frac{D}{H}\right)_v^2 \cong \left(\frac{D}{H}\right)_w^2 \left(\frac{P_{D_2O}^{0}}{P_{H_2O}^{0}}\right) \qquad (3.14)$$

which, again, is consistent with equation 2.9. It directly follows that

Table 3.2 Selected Isotopic Fractionation Factors for the H_2O system

Form: $1000 \ln \alpha_{A-B}^{eq} = C_1 + 10^3 C_2/T + 10^6 C_3/T^2$

Type	Phase A	Phase B	C_1	C_2	C_3	T (K) range	Ref.
D/H	Water	Water vapor	52.612	−76.248	24.844	273–373	1
D/H	Water	Water vapor	−100.0		15.013	258–273	2
$^{18}O/^{16}O$	Water	Water vapor	−2.0667	−0.4156	1.137	273–373	1
D/H	Ice	Water vapor	−94.5		16.289	233–273	2
$^{18}O/^{16}O$	Ice	Water vapor	−28.224	11.839		240–273	3

References: (1) Majoube, 1971; (2) Merlivat and Nief, 1967; (3) Majoube, 1970.

$$\alpha_{w-v}^{D} \cong \sqrt{\frac{P_{H_2O}^0}{P_{D_2O}^0}} \qquad\qquad (3.15)$$

The latter result indicates that the hydrogen isotopic fractionation factor is given by the square root of the ratio of the vapor pressures of H_2O and D_2O. As was shown in chapter 2, this result is consistent with the rule of the mean (equation 2.70). However, this result is not as accurate as equation 3.12, owing to the character of the assumptions made in equation 3.10.

3.3.4 Liquid–Vapor Fractionation in a Closed System

As an example of the utility of the above relationships, one can model the formation of a cloud from an atmospheric air mass. This is a realistic meteorological problem. While water vapor is ubiquitous but invisible, the visible white material in most clouds is small, suspended droplets of liquid water, that has condensed from the ambient atmospheric vapor. The distribution of isotopes among the two phases can, in an ideal closed system, be quantitatively described by combining a statement of isotopic equilibrium with a mathematical statement of material balance that reflects the conservation of the different isotopic species as they are redistributed among the water and vapor phases.

For a closed system, the relevant statement of material balance is (equation 1.13):

$$R_{\text{system}} = X_v R_v + X_w R_w \qquad\qquad (3.16)$$

where R_{system}, R_v, and R_w refer to either the D/H or the $^{18}O/^{16}O$ ratios of the entire system and its constituent vapor and liquid water drops, respectively. Because there are only two phases, the mole fractions of vapor and liquid, X_v and X_w, must sum to unity. The statement of isotopic equilibrium is simply

$$\alpha = R_w/R_v \qquad\qquad (3.17)$$

These coupled relationships directly lead to the equations

$$R_w = \frac{\alpha R_{\text{system}}}{\alpha - \alpha X_v + X_v} \qquad\qquad (3.18a)$$

and

$$R_v = \frac{R_{\text{system}}}{\alpha - \alpha X_v + X_v} \qquad\qquad (3.18b)$$

These relations express the isotopic ratios of the liquid or vapor in terms of the isotopic ratio of the overall system and the mass fraction of the system that is in the vapor phase. For a system that initially was composed of 100% vapor, then R_{system} is identical to the isotopic ratio of that initial vapor, and X_v represents the fraction f of that vapor that remains after some condensation has occurred. These equations may be directly converted to the δ-notation; for example,

$$\delta_w = \frac{\alpha \delta_{system} + 1000f(\alpha - 1)}{\alpha - \alpha f + f} \tag{3.19}$$

On a graph of the δ-values versus f, this expression is almost perfectly linear (Figure 3.5). In fact, because α is close to unity, the approximate linear relationship is

$$\delta_w \cong \delta_{system} + 1000f(\alpha - 1) \tag{3.20a}$$

while the linear approximation for the vapor is

$$\delta_v \cong \delta_w - \Delta_{w-v} \tag{3.20b}$$

where the constant Δ_{w-v} is approximately equal to $1000(\alpha - 1)$. Of course, δ_v is more precisely given by the relationship

$$\alpha_{w-v} = \left(\frac{1000 + \delta_w}{1000 + \delta_v}\right) \tag{3.21}$$

which may be used in conjunction with equation 3.20a, or, if more accuracy is needed, with equation 3.19.

It is useful to consider the trend that the condensate would make on a standard graph of the δD values versus $\delta^{18}O$. This may be found by first taking the partial derivative of equation 3.20a with respect to f:

$$\partial \delta_w / \partial f \cong 1000(\alpha - 1) \tag{3.22}$$

Of course, equations of the form of equation 3.22 may be written for both hydrogen isotopes and oxygen isotopes. The desired slope is then found by simply dividing those two relationships. The result is

$$\text{Slope} = \partial \delta D_w / \partial \delta^{18}O_w \cong (\alpha^D_{w-v} - 1)/(\alpha^{18}_{w-v} - 1) \tag{3.23}$$

Because the deuterium fractionation factor at 25°C is approximately 1.0794, while that for oxygen is 1.0094, then the slope of the trend indicated by equation 3.23 will be close to 8.4. Considering all the simplifications of this model, this value is quite close to the observed slope of the MWL. It is indeed likely that the process of equilibrium atmospheric condensation is key to the formation of meteoric waters, as would be anticipated. However, even if the original mass of atmospheric vapor is completely condensed, the entire range of δD and $\delta^{18}O$ values that can be produced in the water by this process is only about 79 and 9 per mil, respectively, For comparison, the range of these values observed among the meteoric waters of Earth is about seven times larger than this (Figure 3.1). Additional processes must therefore take part in the formation of meteoric waters.

3.4 Rayleigh Fractionation

Rayleigh fractionation is an important, open system process that involves the progressive removal of a fractional increment of a trace substance from a larger reservoir. A consistent relationship, such as a distribution coefficient, equilibrium constant, or a fractionation factor, is maintained between the reservoir and each increment at the instant of its formation, but, once formed, each increment is thereafter removed or otherwise isolated from the system. The mathematical constraints of this process include the distribution coefficient along with a statement of material balance in the shrinking system. These constraints may be combined into a single differential equation that may be integrated to a well-known relationship discovered by Rayleigh (1902). Examples of this common process include condensation, distillation, and the formation of crystals from a melt or a solution. As shown in the following, Rayleigh fractionation explains many different characteristics of meteoric waters, and provides a first-order explanation of the MWL.

3.4.1 Differential Equation for a Rayleigh Process

In the following development of the Rayleigh differential equation, the point of view will be adopted of an environment of progressive condensation of water vapor, using oxygen isotopes as the germane example. Nevertheless, the identical development can be made much more general by (1) substituting N^*, representing the trace element or isotope, for ^{18}O in the following; (2) by substituting N, representing the more abundant element or isotope, for ^{16}O; and (3) recognizing that any distribution coefficient may be used in place of the liquid–vapor isotopic fractionation factor.

For the condensation of atmospheric water vapor, the key constraint is the equilibrium condition

$$\left(\frac{^{18}O}{^{16}O}\right)_w = \alpha \left(\frac{^{18}O}{^{16}O}\right)_v \tag{3.24}$$

where α refers to the liquid–vapor fractionation factor for oxygen. Because, under Rayleigh assumptions, the tiny amount of liquid formed at any time is immediately removed from the system, in this case as meteoric precipitation, this constraint may be converted to a differential equation:

$$\frac{d^{18}O}{d^{16}O} = \alpha \left(\frac{^{18}O}{^{16}O}\right)_v \tag{3.25a}$$

The preceding step is deceptively simple. Here, the $^{18}O/^{16}O$ ratio of the condensed water has been expressed in terms of the incremental loss of oxygen from the large reservoir of vapor that constitutes the system. In this manner, mass balance effects have been automatically incorporated into the equilibrium condition, and the result has been converted into a differential equation that can be directly solved. An essential and easily overlooked point is that the variables in equation 3.25a now refer exclusively to the vapor phase.

While equation 3.25a may be directly integrated, it is very convenient to first rewrite it in the following form:

$$\frac{d^{18}O}{^{18}O} - \frac{d^{16}O}{^{16}O} = (\alpha - 1)\left(\frac{d^{16}O}{^{16}O}\right)_v \tag{3.25b}$$

Equation 3.25b may be simplified by introducing the progress variable f, defined as the fraction of original material remaining in the system at any time. In a Rayleigh problem, the variable f decreases from a value of unity at the beginning of the problem to zero at its final mathematical limit. For the particular case of oxygen isotopes, where the subscript letters i refer to the amount initially present, f is

$$f = \frac{{}^{18}O + {}^{17}O + {}^{16}O}{{}^{18}O_i + {}^{17}O_i + {}^{16}O_i} \qquad (3.26a)$$

Because ${}^{18}O$ and ${}^{17}O$ are present only in trace quantities, and, also, because the fractionation factors are close to unity, f is very nearly equal to the following ratio:

$$f \cong \frac{{}^{16}O}{{}^{16}O_i} \qquad (3.26b)$$

This ratio very closely approximates the total number of moles of water vapor in the system normalized to the number of moles at the beginning of the process. In this particular case, $d \ln f$ is identical to $d \ln {}^{16}O$, which may be established by taking the logarithm of equation 3.26b and then differentiating:

$$d \ln f \cong d \ln {}^{16}O - d \ln {}^{16}O_i = d \ln {}^{16}O \qquad (3.26c)$$

because the derivative of ${}^{16}O_i$, a constant, is zero.

Finally, note that the left-hand side of equation 3.25b is identical to $d \ln R$, where R is the ${}^{18}O/{}^{16}O$ ratio of the reservoir, which in this particular case is the vapor. Substituting $d \ln f$ for $d\ {}^{16}O/{}^{16}O$ gives the key differential equation for a Rayleigh process:

$$\boxed{d \ln R = (\alpha - 1)d \ln f} \qquad (3.27)$$

Equation 3.27 is more general than are any of its various integrated forms, and, in complex problems, it provides the appropriate starting point for numerical integration. The equation may be applied to many problems in physical chemistry by recognizing that R may be generalized to the ratio N^*/N of the reservoir, f to the ratio N/N_i, and α to any distribution coefficient, even if the partition coefficient is variable or if it is not an equilibrium one.

3.4.2 Rayleigh Equation for a Constant Fractionation Factor

In the case where the fractionation factor (or distribution coefficient) is constant, equation 3.27 may be directly integrated. The well-known result was originally given by Rayleigh (1902):

$$\boxed{\left(\frac{R}{R_i}\right) = f^{\alpha-1}} \qquad (3.28)$$

where R_i represents the ratio of interest for the reservoir at the beginning of the process, when $f = 1$. Because α has been assumed to be a constant, it makes no difference in this case whether R refers to the liquid or to the vapor phase. Thus, converting to the δ-notation,

$$\boxed{\left(\frac{1000 + \delta_w}{1000 + \delta_{w,i}}\right) = f^{\alpha_{w-v}-1}}$$

$$(3.29a)$$

for the water, and

$$\left(\frac{1000 + \delta_v}{1000 + \delta_{v,i}}\right) = f^{\alpha_{w-v}-1}$$

$$(3.29b)$$

for the vapor, where, again, at all times,

$$\alpha_{w-v} = \left(\frac{1000 + \delta_w}{1000 + \delta_v}\right)$$

$$(3.29c)$$

In many elementary treatments, the Rayleigh equation for isotope problems is given as:

$$\delta_w = 1000(f^{\alpha-1} - 1)$$

$$(3.29d)$$

The last equation is correct only for the case where α is constant *and* $\delta_{w,i}$ equals zero. For problems where the latter is not appropriate, equation 3.29a should be used. In such a case, $\delta_{w,i}$ may be easily found from the isotopic ratio of the initial system, $\delta_{v,i}$, using equation 3.29c.

As an example, Figure 3.5 compares the results for closed system cloud formation (equation 3.19) with the open system, "continuous precipitation" scenario that can be modeled with equation 3.29a. Note that the isotopic depletions achievable in a Rayleigh process are much more extreme than can be afforded in a closed system. For example, in the extreme limit where f approaches zero, then R approaches zero and the δ-values approach $-1000‰$, indicating complete removal of the heavy isotope from the vapor phase.

Systems that involve water are of particular interest, and it is commonly convenient to directly compare oxygen and hydrogen isotope variations. In this case, because f will be virtually identical for both isotopic systems, the Rayleigh model for a constant fractionation factor becomes

$$\left(\frac{1000 + \delta D}{1000 + \delta D_i}\right) = \left(\frac{1000 + \delta^{18}O}{1000 + \delta^{18}O_i}\right)^{(\alpha_D-1)/(\alpha_{18}-1)}$$

$$(3.30a)$$

On a graph of δD versus $\delta^{18}O$ values, a Rayleigh process will produce a curve that extends from the initial point $(\delta D_i, \delta^{18}O_i)$ to the point $(-1000, -1000)$ in the limit at $f = 0$ (Figure 3.6). The instantaneous slope of this relationship is:

$$\boxed{\frac{d\,\delta D}{d\,\delta^{18}O} = \left(\frac{\alpha_D - 1}{\alpha_{18} - 1}\right)\left(\frac{1000 + \delta D}{1000 + \delta^{18}O}\right)}$$

$$(3.30b)$$

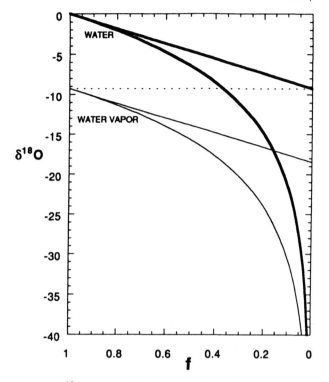

Figure 3.5 Graph of $\delta^{18}O$ values of liquid water (heavy lines) and water vapor (light lines) versus the fraction f of water vapor remaining in closed and open systems. The closed system trends are almost perfectly linear, while the open system (Rayleigh) trends are strongly curved and project to a value of -1000 at $f = 0$. The light dotted line represents the δ-value of the initial vapor.

3.4.3 Rayleigh Equation for a Large Concentration of the Minor Isotope

The Rayleigh equation must be modified for cases where the concentration of the minor isotope is substantial, as it would be for highly enriched waters or for certain other isotopic systems. For an element primarily composed of two isotopes, N and N^*, equation 3.26 may be generalized to

$$f = \frac{N(1+R)}{N_i(1+R_i)} \tag{3.31a}$$

where R is the ratio N^*/N and the subscript i indicates the initial value. This equation may be differentiated to give

$$d\ln f = d\ln N + d\ln(1+R) \tag{3.31b}$$

Thus, the Rayleigh differential equation becomes (R. E. C., unpublished):

$$d\ln R = (\alpha - 1)\{d\ln f - d\ln(1+R)\} \tag{3.31c}$$

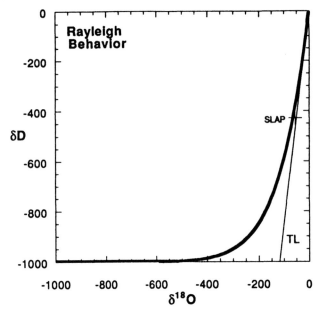

Figure 3.6 Graph of δD versus δ^{18}O values for a constant alpha Rayleigh process, extended all the way to the limiting point $(-1000, -1000)$ at $f = 0$. The assumed fractionation factors of 1.00937 for oxygen and 1.07935 for hydrogen are appropriate for water–water vapor equilibrium at 25°C. The slope of tangent line TL is 8.47. Practically all meteoric waters on Earth lie between the upper right-hand corner and the cross that represents SLAP; f becomes vanishingly small beyond this point.

This equation may be numerically integrated for any case of interest. For the case where the fractionation factor (or distribution coefficient) is constant, it may be directly integrated:

$$\frac{R}{R_i} = \left(\frac{(1 + R_i)f}{1 + R}\right)^{\alpha - 1}$$

(3.31d)

In effect, the above expressions correct the quantity f to include the content of the minor isotope. These equations all reduce to the expressions in the previous sections if the amount of the trace isotope is very small.

3.4.4 Rayleigh Equation for a Variable Fractionation Factor

In many realistic situations, the value of the fractionation factor (or distribution coefficient) may itself vary considerably during a Rayleigh process. Formation of meteoric water is a fine example. Atmospheric air masses commonly undergo rainout during progressive cooling; fundamentally, it is the reduction of the vapor pressure of water with declining temperatures that forces meteoric precipitation to occur. As cooling and condensation progress, then α must change because of its dependence on temperature.

For the case where α is variable, equation 3.27 cannot be integrated without independent information concerning its relationship to f. The simplest assumption is that, at any stage, an appropriate average value for α may be taken at some mean stage of the continuous process; that is, at a stage intermediate between the beginning ($f = 1$) and the stage in question. It may be argued that at a given stage f, the appropriate average fractionation factor would be that which occurred at stage $\sqrt{f}$. Dansgaard (1964) has alternatively argued that the appropriate factor for any given stage of a cooling process is that which occurred at the mean temperature along the path to that stage. In either case, the mean value theorem of calculus may be used to approximate equation 3.27 by the following:

$$d \ln R_v \cong (\alpha_M - 1) \, d \ln f \tag{3.32}$$

Following Dansgaard's argument, for the above equation and its solutions, there are three relevant temperatures. First, at the beginning of the process where f equals unity, the temperature is T_i and the relevant fractionation factor is α_i appropriate for that temperature. Second, at the stage in question, characterized by some value of variable f and the instantaneous temperature T, the fractionation factor is the value of α appropriate for that temperature. Finally, for the average condition of the process, the mean temperature T_M is simply $(T + T_i)/2$, and the hypothetical fractionation factor α_M in equation 3.32 is simply the fractionation factor at that mean temperature.

With the above assumptions, and using the mean value theorem, the approximate solution to equation 3.32 is, for the vapor phase,

$$\left(\frac{R}{R_i}\right)_v \cong f^{\alpha_M - 1} \tag{3.33a}$$

Interestingly, because α is variable, the equation for the condensed water has an additional but minor complication, and becomes

$$\boxed{\left(\frac{R}{R_i}\right)_w \cong \frac{\alpha}{\alpha_i} f^{\alpha_M - 1}} \tag{3.33b}$$

If these relationships are converted to the δ-notation, equation 3.25a for the vapor becomes

$$\left(\frac{1000 + \delta_v}{1000 + \delta_{v,i}}\right) = f^{\alpha_M - 1} \tag{3.33c}$$

which is identical to equation 3.29b, except that α was replaced by α_M. The equation for the water is somewhat different:

$$\boxed{\frac{1000 + \delta_w}{1000 + \delta_{w,i}} \cong \frac{\alpha}{\alpha_i} f^{\alpha_M - 1}} \tag{3.33d}$$

In the general case, no matter how complicated the variations of α with f, simple iterative methods may be used to numerically evaluate differential equation 3.27. Basically, this is accomplished by changing the value of f incrementally, computing the relevant value of α, and then using the differential equation to define the incremental change in R. At each step, that difference is then added to the preceding value of R. Successive repetition of this calculation will generate the desired numerical solution.

3.4.5 Application of the Rayleigh Equation to Meteoric Waters

It is useful to use the Rayleigh equation to model the meteoric precipitation on Earth. As will be seen, many features of meteoric precipitation may be explained by this model, particularly considering its inherent simplicity compared with the complex workings of the enormous hydrologic cycle.

A Rayleigh model requires a starting reservoir, in this case one of water vapor. Most meteoric precipitation on Earth originates from evaporation from the ocean in the trade wind belt, centered on 30° N and S latitudes; this is the only large belt on Earth where evaporation consistently exceeds precipitation. Marine vapors in this belt typically have $\delta^{18}O$ values of about $-12.0‰$, and δD values of about $-86‰$; moreover, the average humidity is about 75% while the surface temperature is approximately 20°C (e.g., Craig and Gordon, 1965). Because the humidity is not 100%, this vapor is in neither chemical nor isotopic equilibrium with ocean water. The fact that the $\delta^{18}O$ and δD values of this vapor lie very close to the trace of the MWL is very interesting and will be discussed more fully in chapter 4.

Next, a model for the variation in f is required. For a cooling system, a simple approximation for f is given by the saturation vapor pressure (P_{sat}) of ordinary water at the temperature of the given stage, divided by the saturation vapor pressure (P^0_{sat}) at the dew point temperature where condensation first began. Thus,

$$f \cong (P_{sat})/(P^0_{sat}) \tag{3.34a}$$

A more accurate value for f is given by the ratio of the quotient of the saturation mixing ratios at the particular stage (r_{sat}) and the value (r^0_{sat}) at the dew point ($f = 1$) stage:

$$\boxed{f = (r_{sat})/(r^0_{sat})} \tag{3.34b}$$

The latter expression correctly normalizes the air mass to a constant amount of dry air. It also allows for changes in the total atmospheric pressure, P_{tot}, as well as temperature during the cooling process, depending on the particular atmospheric conditions. The most straightforward model is for isobaric cooling, where P_{tot} remains constant. A more realistic model, termed "pseudoadiabatic," allows for the progressive reduction of total pressure during cooling, with the $P–T$ trajectory of the air mass remaining saturated in water vapor at all times. Such a saturated adiabatic trend differs from a simple adiabatic trend in that the latent heat released by the condensation of vapor is taken into account.

For the given example pertaining to the trade wind belt, the reference temperature where $f = 1$ may be determined. The dew point of a 20°C air mass with a relative humidity of 75% is 15.4°C. For simplicity here, a starting temperature of 15°C will be employed, at which the saturation vapor pressure is 12.788 torr and the $\delta^{18}O$ and δD values of this vapor are as given above. Moreover, the $\delta^{18}O$ and δD values of the first, microscopic droplet of water that forms from this vapor may be computed from the fractionation factors at 15°C (Table 3.2); respectively, these values are −1.9 and −2.7‰. Note that this value is different than seawater, and lies quite close to the trace of the MWL.

From this point forward, everything is in place to complete the Rayleigh calculation. The atmospheric vapor becomes progressively depleted in the heavy isotopes, because, at every stage, the condensed water that is removed from the system has higher D/H and $^{18}O/^{16}O$ ratios than the remaining vapor. A set of numerical calculations (equation 3.27) for isobaric and pseudoadiabatic cooling models is provided in Table 3.3. The condensation of liquid water is considered to proceed to temperatures well below 0°C, which was achieved by simply extrapolating the equations for the saturation vapor pressure for water and for the water–water vapor fractionation factors to subzero temperatures. This nonequilibrium condition is realistic, because many observed water clouds are supercooled by tens of degrees.

The slope column in Table 3.3 refers to the instantaneous slope at each point along the curve, according to the numerical calculation. The approximate value of this slope may be determined from equation 3.32, if relations of this type for the hydrogen and oxygen isotopes are divided. This gives for the approximate slope,

$$\frac{d\,\delta D}{d\,\delta^{18}O} \simeq \frac{(\alpha_M^D - 1)}{(\alpha_M^{ox} - 1)} \frac{1000 + \delta D}{1000 + \delta^{18}O} \tag{3.35}$$

Note that the true instantaneous slope for the vapor trend, even for a variable fractionation factor, is still given by equation 3.30b. The slope equation for the

Table 3.3 Rayleigh Condensation Models for Water–Water Vapor

T (°C)	P_{total}	Pseudoadiabatic Cooling Model				Isobaric Cooling ($P_{total} = 1$ atm)			
		f	$\delta^{18}O$	δD	Slope	f	$\delta^{18}O$	δD	Slope
15.0	755	1.0000	−1.88	−2.7	7.86	1.0000	−1.88	−2.7	8.28
10.0	674	0.8040	−3.69	−17.0	8.04	0.7170	−4.89	−27.7	8.34
5.0	601	0.6408	−5.68	−33.2	8.21	0.5087	−8.14	−54.8	8.37
0.0	536	0.5039	−7.90	−51.6	8.36	0.3568	−11.7	−84.2	8.38
−5.0	480	0.3900	−10.4	−72.7	8.48	0.2472	−15.5	−116.1	8.36
−10.0	432	0.2960	−13.3	−97.0	8.57	0.1690	−19.6	−150.4	8.30
−15.0	393	0.2193	−16.5	−125.3	8.62	0.1139	−24.1	−187.4	8.20
−20.0	362	0.1579	−20.4	−158.2	8.61	0.0756	−28.9	−227.0	8.06
−25.0	340	0.1098	−24.8	−196.5	8.52	0.0493	−34.2	−296.2	7.87

condensed water is complicated, but the above equation is a useful approximation for it.

As shown by the numerical calculations in Table 3.3, the slope of a pseudoadiabatic trend on a standard δD versus $\delta^{18}O$ graph is close to the value of 8 observed for the MWL (Figure 3.7). It is also evident that this slope is not perfectly uniform, and that the calculated trend is slightly curved rather than linear. It is, in fact, impossible for a Rayleigh curve to be perfectly linear, because it must ultimately project to the point $(-1000, -1000)$, as illustrated for the simple example in Figure 3.6. The MWL likewise cannot be perfectly linear. For example, the standard SLAP (standard light Antarctic precipitation) has very-well-characterized δD versus $\delta^{18}O$ values of -55.5 and -428, respectively. Note that these values do not lie precisely along the linear, slope 8 trend of equation 3.1, suggesting that the actual MWL exhibits a slight, concave upward curvature similar to that shown in Figure 3.6.

Another matter of particular interest is the correlation of the δD versus $\delta^{18}O$ values of meteoric precipitation with temperature. The calculations in Table 3.3 are compared with Dansgaard's trend in Figure 3.8. Clearly, the Rayleigh model is not only capable of producing much larger isotopic depletions than is possible in a closed system, but also the correspondence between those depletions and the temperature is similar to what is observed in nature.

Several improvements to the above model may be made. First, at lower temperatures, the effects of condensation of ice from the vapor could be incorporated (see below); this requires use of the vapor pressure and isotopic fractionation relationships appropriate for ice–vapor. Second, more realistic atmospheric mod-

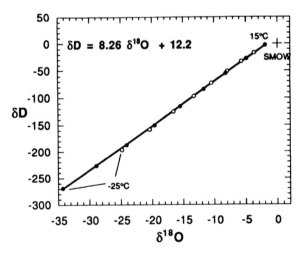

Figure 3.7 Graph of calculated δD and $\delta^{18}O$ values for a Rayleigh model, assuming pseudoadiabatic (open circles) and isobaric (closed circles) cooling following water saturation at 15°C. The calculation provides a first-order simulation for the slope of the MWL; the indicated regression line is for the isobaric model. The cross indicates SMOW. Numerical data are from Table 3.3.

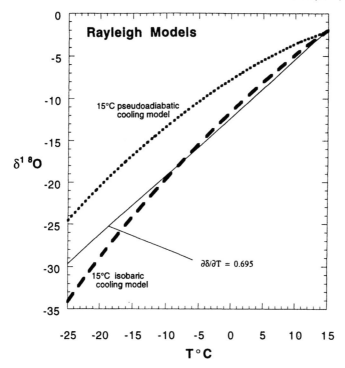

Figure 3.8 Calculated relationship between the $\delta^{18}O$ values of meteoric precipitation and temperature for a Rayleigh process in cooling, water-saturated air masses with initial dew points of 15°C. The curves for pseudoadiabatic and isobaric cooling are compared with the slope of 0.695 that approximates observed changes in $\delta^{18}O$ values with temperature in temperate zones. Numerical data are from Table 3.3.

els may be used to calculate f. Many other refinements are possible, but the major elements of this physical problem are illustrated by the example provided.

The various successes and shortcomings of a Rayleigh description of Earth's complex hydrologic cycle may be summarized as follows.

Rayleigh model explains or approximately explains:
1. The origination of the MWL near ocean water,
2. The slope of the MWL,
3. The large depletions in δD and $\delta^{18}O$ observed in meteoric precipitation in cold regions,
4. The temperature dependence of the δD and $\delta^{18}O$ values of meteoric precipitation,
5. The high global average of the δD and $\delta^{18}O$ values in meteoric precipitation,
6. The correlation of δD and $\delta^{18}O$ values with the amount of precipitation, particularly the existence of large amounts of isotopically heavy precipitation in the tropics and comparatively small amounts of highly depleted precipitation near the poles.

Rayleigh model does not explain or explicitly account for:

1. The "deuterium excess": that is, the y-intercept of the MWL,
2. The detailed global average of the δD and $\delta^{18}O$ values for precipitation of about $(-4.5, -26)$,
3. The existence of clouds: that is, condensed water droplets that are *not* removed instantaneously,
4. The continual contributions of evapotranspiration to real air masses,
5. Atmospheric mixing effects,
6. The presence and thermodynamic behavior of supercooled water in clouds,
7. Meteorological details.

3.5 Examples of Natural Meteoric Precipitation

3.5.1 Antarctic Precipitation

Of all the continents, Antarctica best illustrates the effects of meteorological parameters on the D and ^{18}O contents of natural precipitation. This nearly circular continent, approximately centered on the South Pole, is completely surrounded by open ocean with a surface temperature close to 0°C. Antarctica is meteorologically isolated not only by its geographic location, but also by the circumferential circulation of the circum-Antarctic current, as well as by the polar atmospheric cell that features rising air in the low-pressure belt near the coast, advection of the air toward the high-pressure polar cap, then descent and return of the air, now nearly completely devoid of moisture, toward the coast. Moreover, the continent is characterized by extreme temperature variations. Conditions become increasingly severe toward the center of the continent; for example, at the South Pole station, altitude has increased to 2.8 km, surface pressure has decreased to about 510 torr (680 mb), and mean annual surface temperature has dropped nearly to −50°C.

A useful, first-order isotopic model for Antarctic precipitation may be crafted with the Rayleigh model. Water vapor derived from the Antarctic Ocean is introduced into an unsaturated atmosphere. Condensation is assumed to begin upon cooling this air mass to 0°C. The parameter f is unity at this point, and for subsequent stages f may be determined from equations 3.6 and 3.34b. In the simplest case, both f and α are determined from the appropriate relationships for the ice–vapor system. The δ-values for the original vapor reservoir, and hence the 0°C condensate, were arbitrarily chosen as an initial condition to provide reasonable correspondence between the model and the δ-values and temperatures observed at coastal localities.

Numerical calculations for Rayleigh condensation along the pressure–temperature paths defined by two atmospheric models, isobaric cooling and a pseudoadiabat, are presented in Table 3.4. Isobaric cooling is not a realistic overall physical model for Antarctica, because the average pressure at the polar cap is only about 490 torr (650 mb). However, the more-realistic pseudoadiabatic model provides a reasonably good fit for Antarctic precipitation (Figures 3.9 and 3.10). In particular, the model closely reproduces the trend of the MWL, although

Table 3.4 Rayleigh Condensation Models for Ice–Vapor

		Pseudoadiabatic Cooling Model				Isobaric Cooling ($P_{total} = 1$ atm)			
T (°C)	P_{total}	f	$\delta^{18}O$	δD	Slope	f	$\delta^{18}O$	δD	Slope
0	746	1.0000	−7.1	−47.0	8.05	1.0000	−7.1	−47.0	8.11
−5	684	0.7158	−11.5	−82.0	7.93	0.6559	−12.8	−92.8	7.90
−10	627	0.5046	−16.3	−120.1	7.77	0.4237	−19.1	−141.5	7.66
−15	574	0.3499	−21.7	−161.4	7.59	0.2693	−25.9	−192.8	7.38
−20	527	0.2381	−27.7	−206.0	7.37	0.1681	−33.4	−246.6	7.07
−25	485	0.1587	−34.3	−254.0	7.11	0.1030	−41.5	−302.7	6.73
−30	447	0.1033	−41.7	−305.3	6.81	0.0619	−50.4	−360.6	6.35
−35	414	0.0655	−49.9	−359.9	6.46	0.0364	−60.0	−419.9	5.94
−40	387	0.0403	−59.1	−417.4	6.07	0.0209	−70.5	−479.9	5.50

deviations appear for the most depleted samples (Figure 3.9). In addition, the model accurately explains the observed correspondence between the $\delta^{18}O$ values and measured cloud temperatures (Figure 3.10). Because large temperature inversions are common in central Antarctica, correlations between the character of precipitation and ground temperatures are less useful.

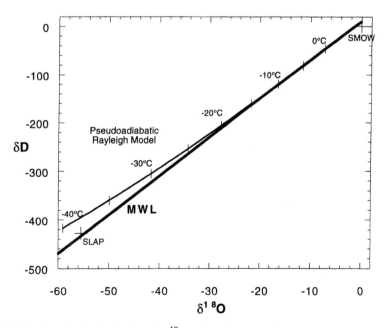

Figure 3.9 Graph of calculated δD and $\delta^{18}O$ values for a Rayleigh model, assuming pseudoadiabatic cooling following ice–vapor saturation at 0°C. The calculation provides a reasonable simulation for Antarctic precipitation, but shows a positive deviation from the MWL at low values. The crosses indicate the SMOW and SLAP international standards. Numerical data are from Table 3.4.

Figure 3.10 Graph of $\delta^{18}O$ values versus T that were observed at several Antarctic stations compared with a Rayleigh model (heavy curve) calculated for cooling along a pseudoadiabatic path that begins at vapor–ice saturation at 0°C. Maximum (▲) and minimum (▼) cloud temperatures, and ground (−) temperatures, are indicated; strong inversions are common at the South Pole station, so ground temperatures tend to deviate well to the lower left. Data are from Aldaz and Deutsch (1967), Epstein et al. (1963), and Picciotto et al. (1960). Numerical data are from Table 3.4.

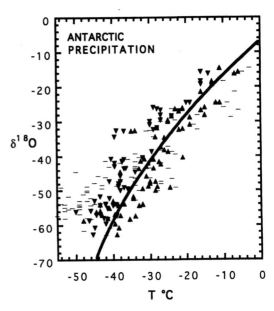

3.5.2 Ice Cores

The great ice sheets of Greenland and Antarctica are several kilometers thick and preserve a direct and sequential record of "fossil" meteoric precipitation. Studies of the Greenland ice sheet first established the preservation of ice that is many tens of thousands of years old, dating back to the last glacial maximum and even before. The isotopic record preserved in the ice exhibits a pronounced minimum that coincided with the glacial maximum, as well as many smaller variations that correspond to shorter term climatic fluctuations, with the overall record having a "sawtooth" pattern (e.g., Dansgaard et al., 1969). Subsequent study of a deep core from Byrd Station, Antarctica revealed similar variations in the $\delta^{18}O$ values of ice, from about −32‰ today to about −40‰ about 17 ka ago (Epstein et al., 1970). Moreover, the precipitation from the glacial maximum did not lie along the present-day MWL, but along a "fossil" MWL that had a deuterium excess close to zero (R. E. C., unpublished, 1974). Interestingly, these two records demonstrated that the glacial maximum in the southern hemisphere approximately coincided with that in the northern hemisphere (Epstein et al., 1970), dooming climate models that predicted that the ice ages would be out of phase between the hemispheres.

Of all the ice cores, however, that from the Vostok Station, Antarctica, located deep in the continental interior at 78°28′S, 106°48′E at an elevation of 3488 m, is most impressive (Lorius et al., 1985). The current mean annual temperature at this station is only −55.5°C, and the vapor pressure of ice is so low that the station receives only 2.2–2.5 g/cm^2 of precipitation every year. Consequently, over the last century, the average δD value of ice at this station has been only −438‰ (Jouzel et al., 1987). More remarkable is the fact that the ice sheet beneath Vostok Station is 3.7 km thick, and deep ice cores recovered there contain fossil ice not only from the last ice age near 17 ka, but also from several previous glacial periods (Petit et al., 1997).

The δD values of the Vostok core conform to several "sawtooth-shaped" climatic cycles (Figure 3.11). Specifically, ice that represents the glacial maxima is lower, by about 50‰ in δD and by 6‰ in $\delta^{18}O$, than modern precipitation at this locality, suggesting that the temperatures were about 10°C colder during the glaciations (Lorius et al., 1985; Barnola et al., 1987). These "sawtooth" cycles correspond to the last few of the many, rather similar, 100-ka Quaternary cycles known from the $\delta^{18}O$ record of carbonate from deep-sea cores, and are thought to represent a combination of climatic change and changes in the $\delta^{18}O$ value of

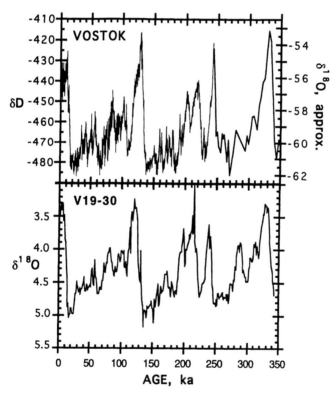

Figure 3.11 Comparison of climate records from equatorial carbonate core V19-30 and from the deep ice core at Vostok, Antarctica. In the upper graph, the measured δD values in the glacial ice (Petit et al., 1997, left scale) are approximately translated to $\delta^{18}O$ values (right scale) using equation 3.1 for the MWL. For comparison, the lower graph shows measured $\delta^{18}O$ values of the carbonate tests of the benthic foram *Uvigerina senticosta* for an equatorial deep-sea core, taken at 3°21'S and 83°21'W (Shackleton and Pisias, 1985); these isotopic variations are similar to those at Vostok, yet smaller and opposite in sign. The sawtooth-type variations exhibited by these parameters record global climatic changes that affected Earth's ice volume and continental and oceanic temperatures.

seawater caused by the growth or retreat of ice caps (cf. Figure 2.9; Emiliani and Shackleton, 1974).

As with most geological records, many complexities exist in ice cores that preclude simple interpretations. For example, because ice flows, ice from different depths in a core did not originate at the same location. At Vostok, the deep ice is thought to have originated as much as 200 km away, at higher altitudes. Moreover, the timescale of the ice is difficult to establish, partly because the accumulation rate is not constant over time. Last, as demonstrated above, the δD and $\delta^{18}O$ values of meteoric precipitation are not a simple function of temperature, but rather depend in a complex way on several meteorological parameters.

3.5.3 Meteoric Precipitation, Central USA

Climatic conditions in continental interiors exhibit large seasonal variations. St. Louis, Missouri is a typical locality, and exhibits an almost perfect sinusoidal variation in mean monthly temperatures from $-1.7°C$ in January to $26.7°C$ in July. The amount of precipitation is remarkably steady over the year, with the average for every month lying in the range 7.4 ± 2.8 cm.

The $\delta^{18}O$ values of semimonthly precipitation composites for St. Louis typically vary by more than 14‰ each year, with the highest values representing late summer and the lowest values representing winter (Figure 3.12). While these variations would seemingly be compatible with the annual temperature variations, the correlation between the $\delta^{18}O$ values and the ground temperatures for individual storm events is weak. Moreover, the annual pattern of the $\delta^{18}O$ variations is regular but not sinusoidal. Relatively little ^{18}O variation occurs during most of the year, but sharp depletions occur during the winter months. One important control over these variations may be the position of the jet stream, which generally lies near the "polar front" that separates polar from temperate zone air masses. For most of the year, St. Louis lies well to the south of the polar front, and the isotopic character of meteoric precipitation is rather uniform. During the winter months, however, St. Louis mostly lies north of the polar front, and polar air masses can deposit highly depleted precipitation, with snow from individual storms having $\delta^{18}O$ values as low as $-27‰$.

3.6 Isotopic Variations in Streamflow

3.6.1 Hydrograph Separation

Most rivers and streams exhibit dramatic variations in flow during rainfall events that are correlated with shifts in their oxygen and hydrogen isotope ratios. Traditional hydrologists have emphasized the importance of "overland flow," literally sheet-wash derived from the most recent rainfall, to explain the streamflow variations. Geochemists have developed simple chemical and isotopic methods that show that flow is more commonly dominated by "baseflow" (i.e., "pre-

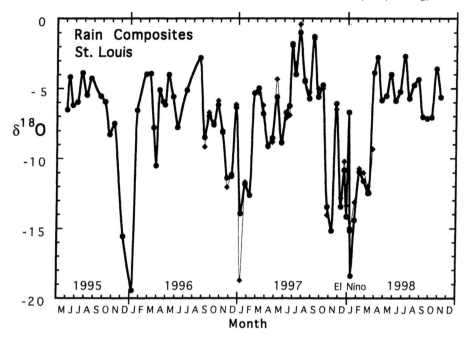

Figure 3.12 Graph of the $\delta^{18}O$ values of semi-monthly composites of meteoric precipitation for two rain gauges in St. Louis, Missouri, notably Ladue (circles) and Washington University (diamonds). For most of the year, the precipitation is within a few per mil of the yearly average $\delta^{18}O$ value of about −6.8. However, sharp negative deviations from this average occur in winter months, showing the influence of polar air masses. The large amount and duration of depleted precipitation in the El Nino winter of 1997–1998 is remarkable and unexpected because temperatures were well above normal for this period.

event water") that is generally derived from groundwater systems (e.g., Sklash and Farvolder, 1979).

The geochemical methods are based on two component mixing relationships. The total streamflow Q_{SF} at any time is assumed to represent the contributions of two endmembers: overland flow Q_{OF} that represents the most recent rainfall, and baseflow Q_{BF} that represents the pre-event water. Thus,

$$Q_{SF} = Q_{OF} + Q_{BF} \qquad (3.36a)$$

The mass transfer of a "conservative" component, such as Cl^-, can be deduced from the appropriate concentrations:

$$Q_{SF}C_{SF} = Q_{OF}C_{OF} + Q_{BF}C_{BF} \qquad (3.36b)$$

or, for the appropriate δD or $\delta^{18}O$ values,

$$Q_{SF}\delta_{SF} = Q_{OF}\delta_{OF} + Q_{BF}\delta_{BF} \qquad (3.36c)$$

For completeness, the relationship for the ratio R of a trace isotope of an arbitrary element whose concentrations are known, would be

$$Q_{SF}C_{SF}R_{SF} = Q_{OF}C_{OF}R_{OF} + Q_{BF}C_{BF}R_{SF} \qquad (3.36d)$$

Note that equation 3.36c is actually a special case of this last relationship, because the chemical concentrations of oxygen and hydrogen are the same in the two endmembers.

The chemical separation method is applied as follows. Dividing equation 3.36b by equation 3.36a gives the familiar form of a binary mixing relationship:

$$C_{SF} = X_{OF} C_{OF} + X_{BF} C_{BF} \tag{3.37a}$$

where the X values represent the mole fractions of water derived from the overland flow and the baseflow, which must sum to unity. All that is required to determine these fractions are measurements of the concentration of a conservative ion in the total streamflow and in the two endmembers. In fact, the concentrations in the overland flow are generally assumed to be negligible because rainwater is dilute, but this assumption is poor because the concentrations rise very rapidly as soon as the rain comes into contact with surface materials, such as clays (R. T. Gregory, pers. comm., 1997). The concentration in the baseflow is generally assumed to be a constant, or identical to that of the streamflow immediately prior to the rain event.

The isotopic separation method is analogously defined by dividing equation 3.36c by equation 3.36a to obtain

$$\boxed{\delta_{SF} = X_{OF} \delta_{OF} + X_{BF} \delta_{BF}} \tag{3.37b}$$

This relationship is basically the same as equation 1.26a for the case of two-component mixing. The mole fractions X_{OF} and X_{BF} must again sum to unity and should be the same as for equation 3.37a. According to the isotopic separation method, these fractions may be determined from measurements of the δ-values of the total streamflow, the baseflow, and the rain event responsible for the particular hydrologic response being investigated.

Ideally, the relative magnitudes of X_{OF} and X_{BF} defined by various chemical and isotopic constituents would all be identical and the same as those defined by hydrogen or oxygen isotopes. Such conformity is not realized in practice. The concentrations of ions increase markedly as soon as rainwater encounters surface materials, even in the overland flow, and the ratios of different ions vary in a complex manner. While the oxygen and hydrogen isotopes are insensitive to such effects, the δ-values of the rainfall may vary with time and location during an individual storm event, resulting in uncertainty in the average values for the precipitation in the watershed. Moreover, the relative sizes of X_{OF} and X_{BF} change with time during a single storm event, with the largest values of X_{OF} being observed on the rising limb of the streamflow peak, and with X_{BF} becoming increasingly more important with time after the peak.

As shown in the following, streamflow is actually composed of a hierarchy of components that have different residence times, and is not well described by two simple endmembers. Moreover, the geochemical character of "baseflow" is not constant over time. Nevertheless, the chemical and isotopic methods of hydrograph separation are superior to the traditional method, and have conclusively demonstrated the importance of "pre-event" water in streamflow. Even the huge

Amazon River is more than 50% baseflow, and this remarkable condition persists even during peak flow conditions.

3.6.2 Floods

Most major rivers in the central USA have undergone serious, repeated spring flooding since the catastrophic summer flood of the Mississippi River during 1993. Following the 163-year flood of 1993, the Mississippi River at St. Louis experienced serious spring floods in 1994, 1995, and 1996, while the lower Missouri River had a destructive flood in 1995. The smaller (10,300 km^2) Meramec River had a very serious late-summer flood in 1993, then even higher levels in spring 1994, and well-above-average spring floods in 1995 and 1996. In 1997, and so far in the El Nino year of 1998, the floods on these rivers have been only slightly above average, with serious flood activity shifting to the north, south, and the west coast of the United States.

The distribution of flooding may not be randomly distributed in time. Serious flooding may increase the probability of floods in the affected areas for 1–3 years. Moreover, the severity of flooding on major rivers, and the height of the flood stages, appears to have increased over the last two centuries. This may be primarily due to the profound engineering modifications of rivers, particularly channelization and levee construction. Moreover, the elimination of wetlands and riparian forests, and the regional destruction of prairies and forests, have profoundly reduced the infiltration capacity of the watersheds.

Oxygen isotope ratios provide a very useful means to define the sources of streamflow. The upper Mississippi, lower Missouri and Meramec rivers near St. Louis have, respectively, average $\delta^{18}O$ values of about −8.5, −10.5, and −6.5 per mil relative to SMOW which reflect the average meteorological conditions of their distinctive watersheds. Significant excursions from these average values can occur during flood periods. During the May 1995 flood, the $\delta^{18}O$ values of all three rivers became practically identical at −6.5 ± 0.2, reflecting greatly increased contributions of surface waters and groundwaters from the central USA (Figure 3.13). It was completely unexpected that the $\delta^{18}O$ values of the largest rivers in the United States would shift by such large amounts, while a small river would shift only little.

Because of this paradox, a more complete data base was obtained for subsequent years. During May 1996, for example, the Mississippi River underwent a modest positive shift to −7.4, while the lower Missouri again shifted all the way to −6.3. As shown in Figure 3.14, the variations in the Missouri River correlate strongly with variations in total volumetric flow, Q, which is normally reported in cfs (cubic feet/second). In contrast, during 1997, when unusually high rainfall shifted to the north, the Mississippi River underwent a large negative excursion to −11.9, while the lower Missouri River was essentially unchanged, reflecting the increased contributions of northern waters that are depleted in heavy isotopes.

Most streamflow is derived from groundwaters, even during peak floods on major rivers. Overland flow is most important on the rising limb of the hydrograph, when the turbidity is highest (10–100× normal) and the water quality is lowest. Quality is somewhat better and turbidity is lower at peak flow, and this

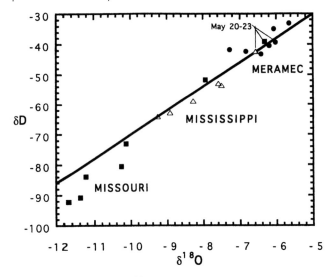

Figure 3.13 Graph of δD versus $\delta^{18}O$ values of the Mississippi (triangles), Missouri (squares), and Meramec (circles) rivers during the 1995 flood year. Normally, these rivers have distinct fields of isotopic variation that reflect the different average character of meteoric precipitation in their watersheds; also, the Missouri River appears to have undergone significant evaporation in upstream reservoirs, causing several samples to shift to the right of the MWL (heavy line). However, during the May 23, 1995 flood peak, the δD versus $\delta^{18}O$ values of all three rivers nearly converged to a single composition, suggesting that streamflow was dominated by displaced groundwaters characteristic of the Missouri area.

condition rapidly improves during the recession period, when streamflow is almost entirely derived from groundwater.

3.6.3 Synthetic Hydrograph

A new mathematical model provides an accurate simulation of streamflow variations following storm events. This model is of interest as it provides the intrinsic time constants of volumetric streamflow that may be compared with the isotopic time constants developed below. Based on analogies with solutions to various diffusion problems, Criss (1997) identified a family of curves that have great semblance to storm hydrographs:

$$\frac{Q}{Q_p} = \left(\frac{eb}{nt}\right)^n e^{-b/t} \tag{3.38}$$

where Q is volumetric discharge, Q_p is peak discharge, t is time, b is the time constant for the basin, and n is a power. This family has many interesting properties. For all such curves, the lag time, representing the interval between the meteoric disturbance at $t = 0$, and the time t of peak flow, where $Q = Q_p$, is given by the ratio b/n. In addition, according to this model, the baseflow recession varies following a storm event as a function of inverse time, not exponentially as

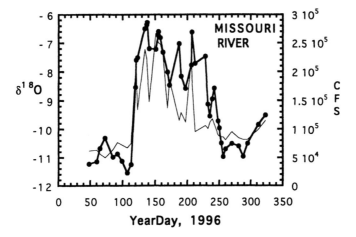

Figure 3.14 Variations in the $\delta^{18}O$ values (heavy line with circles, left scale) and in stream-flow (light line, right scale) for the Missouri River in 1996. Heavy spring flooding in the lower basin produced a large excursion in the $\delta^{18}O'$ values of the water from the normal value of about −11. This deviation gradually became attenuated as streamflow returned to normal in late summer and fall.

argued in standard hydrologic models. Criss (1997) has demonstrated by integrating equation 3.38 over all time that the total amount of streamflow derived from a single rainfall event is finite for almost every choice of n, fulfilling an obvious requirement for a realistic physical model.

A particularly important case of these equations is for the power $n = 3/2$. This case corresponds to forcing by a sharp pulse, or "delta" function, of rainfall. In the above equation, Q becomes analogous to the flux observed over time at a stationary point for a well-known solution of the diffusion equation. For this case, a remarkably simple approximation to equation 3.38 can be made for post-peak flow:

$$Q/Q_p = b/t \qquad (3.39a)$$

This approximation is very accurate following the point where flow is less than 80% of peak flow (i.e., $Q/Q_p < 0.8$). Noting that the time t represents the difference between calendar time t_c and the calendar date (t_{ref}) of the initial forcing (rainfall event), this approximation may be rewritten to give

$$t_c = b(Q_p/Q) + t_{ref} \qquad (3.39b)$$

Now that the model has been defined, it is useful to examine an example. Figure 3.15a shows streamflow variations (solid dots) on the Meramec River from May 14 to June 5 in 1953. This period was chosen because, during the record 1953 drought year, the 1.60 and 0.27 inches of rain that fell, respectively, on May 16 and 17 (yeardays 136 and 137) were followed by several weeks of exceptionally dry weather (US Weather Bureau data). Thus, the 5.4× increase in river flow that resulted from the May 16 storm became attenuated in a regular, virtually undisturbed manner.

As shown on Figure 3.15b, the postpeak flow decreased with time in an almost perfect inverse manner, in agreement with the mathematical model. While the independent and dependent variables on this graph are reversed relative to the normal convention, this is useful because the slope of the linear regression to the data directly gives the time constant b, while the y-intercept indicates the reference time, or storm date. For this gauging station, the time constant is about 2.26 days. Also, the model prediction for the storm date is yearday 136, in agreement with the independent meteorological record. Importantly, only these two parameters— the reference time and the time constant—are needed to regenerate the entire shape of the storm hydrograph, though the observed value for Qp is needed to properly normalize it. For example, if these parameters are introduced into equation 3.38, the entire hydrograph following the May 16 storm is accurately reproduced, as illustrated by the solid curve on Figure 3.15a.

As a rule of thumb, the time constant b increases with increasing average discharge of river basins. While this constant is on the order of 2 days for the

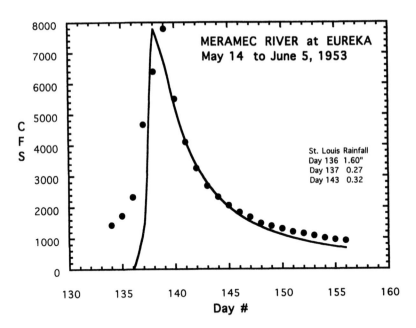

Figure 3.15 (a) Graph of variations in streamflow for the Meramec River in May 1953. Circles are streamflow data from the US Geological Survey gauging station near Eureka. (b) Graph of the variations in streamflow during the baseflow recession period of the May 1953 event shown in part a. Note that the data decrease inversely with time, and not exponentially as generally assumed. The slope of the linear regression line provides the time constant of the flow, while the y-intercept approximates the date of the storm. This intercept agrees with independent weather data from the St. Louis archives, which indicate a storm of 1.6 inches on May 16, 1953. This slope and the y-intercept, along with the observed value of peak flow Q_p, are all that is necessary to generate the entire solid curve shown in part a using equation 3.38.

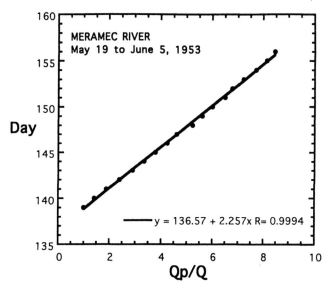

Figure 3.15 (b)

lower Meramec River, it is several weeks for the Missouri and Mississippi rivers near St. Louis.

The same mathematical methods may be used to determine the time constants for different physical, chemical, and isotopic parameters, using appropriate plots. These methods generally reveal that the time constants for variations in these parameters are comparable to that for volumetric flow, though in detail they may shorter or longer. Plots of any given geochemical parameter versus flow thus reveal "hysteresis" loops, with different paths for the rising and falling limbs of the hydrograph. Use of this new method could revolutionize understanding of the transport of various components in streams. As shown in the following, these time constants of volumetric flow and of the geochemical variations are much shorter than the isotopic time constants of the "baseflow" component.

3.6.4 Isotopic Time Constants of Baseflow

An alternate determination of the time constant of streamflow may be made from oxygen isotopes. A new model relates the isotopic response of a given hydrologic system to the isotopic variations and amounts of meteoric input. St. Louis is a fine place to study such effects, because the annual isotopic variations in meteoric precipitation are so large (Figure 3.12).

The $\delta^{18}O$ values of the Meramec River exhibit similar temporal patterns as the local meteoric waters, but the amplitude of the river variations is nearly 10× smaller (Figure 3.16). This result proves that the river is dominated by "baseflow." A simple relationship between the meteorological forcing and the river's isotopic response may be made for a model in which much of the uninterrupted streamflow, that is, the "baseflow," is derived from a large stirred vessel, into which variable amounts of meteoric precipitation (which has variable isotope

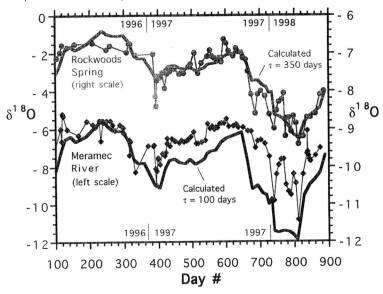

Figure 3.16 Graph of $\delta^{18}O$ variations in Rockwoods Spring (circles; right scale) and the Meramec River (diamonds, left scale) versus time, with day 1 being January 1, 1996. Simulations (heavy lines) of the actual isotopic variations were calculated with equation 3.40 solely using data from the Ladue rain gauge (see Figure 3.12). Note that the amplitude of the isotopic variations in the spring is only 2.5‰, while that for the river is about 6‰ and that for the meteoric precipitation at Ladue exceeds 18‰. This damping results from the long characteristic time constants of 350 days for Rockwoods Spring and 100 days for the Meramec river basin. Data of Frederickson and Criss (1999).

ratios) are periodically added. Most of the variation observed in the river water can thereby be explained by a simple exponential weighting of all pre-existing precipitation events, such that the most recent rainfalls have a greater influence on flow than successively older precipitation events, according to the formulation (Frederickson and Criss, 1999):

$$\delta^{18}O_{river} = \frac{\Sigma\, \delta_j P_j e^{-t_j/\tau}}{\Sigma\, P_j e^{-t_j/\tau}} \qquad (3.40)$$

where δ_j and P_j are the $\delta^{18}O$ value and amount, respectively, for each rain event, and t_j is the time interval between the storm and the river sample. In effect, equation 3.40 *predicts* the $\delta^{18}O$ value of the river or spring from the measured amounts and the $\delta^{18}O$ values of water collected in a nearby rain gauge.

As shown on Figure 3.16, an excellent simulation of the isotopic variations in the Meramec River may be made from rain gauge data, assuming that the time constant τ is close to 100 days. Smaller contributions with $\tau \sim b$ are superimposed, representing the latest storm events. Even better correspondence has been found between the flow of individual springs and rain gauge data, with the springs exhibiting isotopic time constants as long as 350 days (Frederickson and Criss,

1999). An approximate inverse relationship exists between the magnitude of the time constant of the hydrologic system and the amplitude of its annual isotopic variations relative to that of the meteoric input.

This calculation has a several advantages over the standard binary mixing arguments for runoff and baseflow contributions to hydrologic systems. The model demonstrates the dominant contribution of "pre-event" water in the systems, and provides the timescale for mixing and subsurface transport of chemical and isotopic constituents. This method also shows that the "baseflow" component in hydrologic systems has a variable geochemical character. In detail, complex hydrologic systems are probably characterized by a hierarchy of time constants, each represented by a term that has the form of equation 3.40. Nevertheless, it is remarkable how well single time constants describe the isotopic variations in the systems that have been investigated.

The disparity between these isotopic time constants and the time constant of volumetric flow is a fundamental characteristic of hydrologic systems. A likely explanation is that the transmission of pulses in head through groundwater systems is practically instantaneous, while the transport of chemical and isotopic constituents depend on the actual velocity of groundwater flow.

3.7 Isotopic Variations in Groundwaters

3.7.1 Groundwater Sources and Tracing

The isotopic character of groundwater provides unique information on its sources and flowpaths. For example, geographic variations of the $\delta^{18}O$ or δD values of meteoric waters, and especially the marked variations with altitude in mountainous regions, can commonly be exploited to define the source regions of springs. This method has been successfully employed in many regions, including California and Nevada, where many subsurface flow systems that have length scales of tens to hundreds of kilometers have been clearly delineated by different isotopic systems. For example, Rose and Davisson (1996) used ^{13}C and ^{14}C variations in spring waters to discover that a dramatic, 50-km-long "plume" of groundwater, enriched in magmatic CO_2, flows north from the Mt. Lassen volcanic edifice in a pattern consistent with inferences based on the observed variations in the $\delta^{18}O$ values of meteoric precipitation with altitude (Rose et al., 1996).

3.7.2 Isotopic Mapping of Groundwater

Systematic mapping of the isotopic variations in groundwaters can reveal dynamic patterns of flow. As an example, groundwater pumping typically causes reductions in head that alter the natural flow patterns of groundwater. Beneath Sacramento, California, decades of intensive groundwater pumping for domestic consumption have produced two large cones of depression in the water table that are 20 km in diameter and extend far below sea level. These depressions have induced the lateral flow of water from the nearby Sacramento and American rivers, as clearly shown by the oxygen isotope contours in Figure 3.17 (Criss

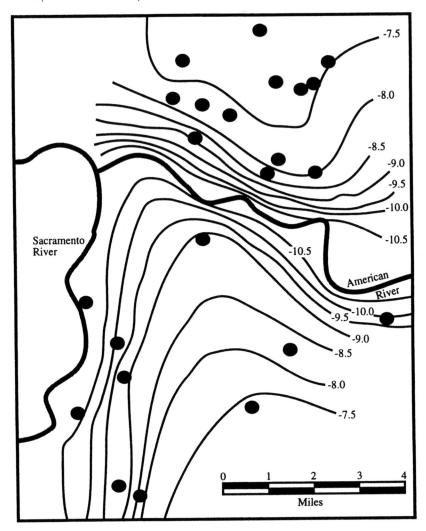

Figure 3.17 Contour map of the $\delta^{18}O$ variations in groundwater beneath Sacramento, California. Sustained pumping has produced two large cones of depression in the water table, inducing the lateral flow of $-11‰$ water from the nearby Sacramento and American rivers. The -9 contour approximates the position where a 50:50 mix of modern river water and Pleistocene groundwater with a value close to -7 is realized. After Criss and Davisson (1996).

and Davisson, 1996). In this case, the isotopic pattern is particularly striking because the $\delta^{18}O$ values of the rivers (~ -11.0) are much lower than those of the local groundwaters (~ -7.0), because the rivers mostly contain "exotic" water derived from high-altitude snowmelt. In addition, the lateral flow system has been amplified by the large areas of impermeable pavement in the city, which impede the downward transport of local meteoric waters that dominated the groundwater system before urban development.

3.7.3 Groundwater Contamination

Processes of groundwater mixing, including contamination, may be revealed by the graphical definition of different geochemical endmembers. As an example, the giant alluvial aquifer in the Central Valley of California, one of the largest and most important aquifers in the world, has become profoundly modified by agricultural practices and overdraft. Isotopic data show that more than 80% of the recharge to this aquifer is now derived from water applied to irrigated fields. The isotopic endmember that represents this modern recharge has several distinctive characteristics, including (1) high ^{14}C and tritium contents indicative of recent surface contact, (2) elevated concentrations of nitrate and sometimes other agricultural chemicals, and (3) elevated δD and $\delta^{18}O$ values that lie well off the MWL due to evaporation during field application (Criss and Davisson, 1996). In this and many other areas, isotopic data can define the endmembers, processes, and rates of groundwater contamination better than any other method.

3.7.4 Basinal Fluids, California Coast Ranges

Isotopic data define two remarkably distinct groundwater types in the California Coast Ranges. The first, normal meteoric groundwaters, occur in shallow, gravity-fed aquifers and commonly emanate from small springs at the base of topographic re-entrants. These Ca-HCO$_3$ waters are distinguished by their relatively dilute character, and by their normal meteoric δD and $\delta^{18}O$ values that decrease regularly with altitude (Figure 3.18a).

The second water type, formation fluid, is characterized by high salinities, major ion chemistry dominated by Na and Cl, and elevated $\delta^{18}O$ (to $+6.3$) and δD values that plot far to the right of the MWL. In this area, these fluids commonly appear as perennial saline springs that can emanate from high elevations, including the very tops of topographic rises, commonly more than 300 m above adjacent valleys. Geochemical characteristics indicate that these waters are formation fluids, and their unusual occurrence indicates that they have become "overpressured"; that is, they have attained pressures significantly greater than hydrostatic, by the active tectonic compression of western California (Davisson et al., 1994). Geochemical thermometers suggest that source temperatures of 30–100°C, corresponding to depths of 1–7 km, are typical (Figure 3.18b).

The distribution of the saline springs shows several relationships to regional tectonic patterns. Notably, the springs generally occur within aseismic belts, correlate with zones where abnormally high pore-fluid pressures have been encountered by drilling, and occur in areas that have experienced significant historical uplift (Melchiorre et al., 1999). The aseismic zones commonly occur in the interiors of fault-bounded structural blocks that are undergoing compression, creating pockets of overpressure that drive the upward flow of the formation fluids (Figure 3.18b).

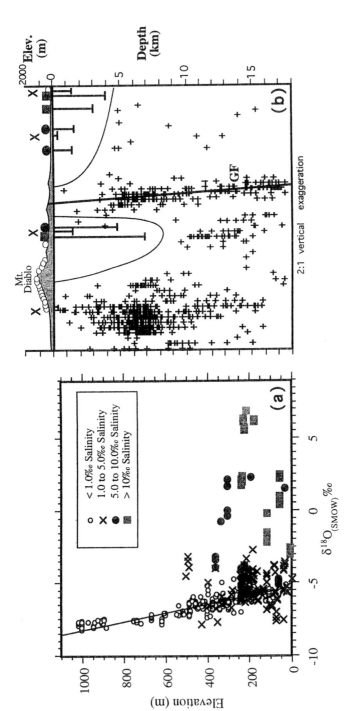

Figure 3.18 (a) Graph of the $\delta^{18}O$ values versus elevation for springs near Mt. Diablo, California. Meteoric springs (open circles) have low salinities and their $\delta^{18}O$ values display a regular, 2.2‰/km decrease with elevation. Formation fluids have high $\delta^{18}O$ values that tend to increase with increasing salinity, but show no correlation with elevation. (b) East–west cross section passing through Mt. Diablo showing springs (symbols as in part a), source depths (vertical lines), and seismic events (crosses). The Mt. Diablo edifice mostly has normal meteoric springs (open circles). The source depths calculated for saline springs lie within aseismic pockets that contain no recent epicenters, suggesting that high pore-fluid pressures promote aseismic deformation at the expense of brittle failure. GF indicates the Grenville Fault. After Melchiorre et al. (1999).

3.8 Summary

Oxygen and hydrogen isotope ratios provide great insight into the processes that govern the hydrosphere and lower atmosphere. The varied origins of many large classes of waters can be distinguished by these intrinsic tracers, which generally behave in a conservative manner in low-temperature environments. Used in combination, the δD and $\delta^{18}O$ values of fluids provide even more powerful signatures of the origins of aqueous fluids and the processes that transform one type into another.

Meteoric waters are arguably the most essential of all resources used by man. It is most fortunate that the δD and $\delta^{18}O$ values of these waters are so variable, yet so regular in their behavior. The remarkable correlation of these values, codified by the empirical meteoric water line, or MWL, allows waters that have a meteoric parentage to be distinguished from practically all other water types on Earth. Moreover, because the geographic variations of the δ-values along the MWL are so large, meteoric waters derived from one locality can commonly be distinguished from waters derived from another. Such differences facilitate the identification of water sources and elucidate mixing effects.

The MWL and associated effects can, to first order, be simulated with a Rayleigh condensation model. Such models have value in quantifying atmospheric processes, and when applied to ice cores they provide insight into the nature of climatic change.

Stable isotope variations have special application to groundwater systems. Their utility in defining the sources of springs or of groundwater components is well established. New models link the isotopic variations in springs and streamflow to the temporally variable meteoric input, and demonstrate the dominance of groundwater in many flowing streams.

3.9 Problems

1. What are the probable origins of the following waters with the indicated characteristics? Suggest a location and occurrence (e.g., London rain) where each sample might have been collected.

Sample	δD	$\delta^{18}O$	Salinity (wt. %)
A	−59	−8.6	0.0
B	−123	−16.6	0.0
C	−97	−6.4	5.7
D	+4	+0.6	3.5
E	−94	−8.7	0.4
F	−428	−55.5	0.0
G	−66	−3.9	0.0
H	−68	+7.3	30.0
I	−39	−6.2	0.0
J	−19	+3.9	19.5

2. The temperatures, δD and $\delta^{18}O$ values, and chloride concentrations obtained by Craig (1966) for fluids from the Salton Sea geothermal system are approximately as indicated below. Suggest an origin for these hot fluids. Use the data to make several quantitative arguments as to whether a significant component of magmatic water is present.

Sample	T (°C)	δD	$\delta^{18}O$	Chlorinity (ppm)
A	17	−77	−9.4	1,500
B	39	−80	−5.5	8,700
C	42	−75	−5.1	12,300
D—brine		−77	+1.4	133,000
D—steam		−76	+0.6	
E	> 300	−75	+3.2	182,000

3. Bohlke and Kistler (1986) report $\delta^{18}O$ values of + 16.4 for mica, + 18.7 for carbonate, and + 21.1 for quartz in a gold-bearing hydrothermal vein from the Irelan Mine in the Alleghany District of the California Mother Lode. Assuming that each pair of minerals represents an equilibrium assemblage, use the fractionation factors for quartz–water, calcite–water, and muscovite–water to estimate both the temperature and the $\delta^{18}O$ value of the hot aqueous fluid that precipitated this vein. How many of these estimates are linearly independent? Are all the minerals in isotopic equilibrium? Suggest an origin for the hydrothermal fluid.

$$1000 \ln \alpha_{qtz-w} = 3.38(10^6/T^2) - 2.90 \qquad \text{(Clayton et al., 1972)}$$
$$1000 \ln \alpha_{cc-w} = 2.78(10^6/T^2) - 2.89 \qquad \text{(O'Neil et al., 1969)}$$
$$1000 \ln \alpha_{musc-w} = 2.38(10^6/T^2) - 3.89 \qquad \text{(O'Neil and Taylor, 1969)}$$

4. Steam separated from a geothermal well at Wairakei, New Zealand is depleted by 3.3‰ in δD and depleted by 2.2 in $\delta^{18}O$ relative to the hot fluid (Bottinga and Craig, 1968). However, at the Salton Sea geothermal system, the steam is enriched in deuterium while still being depleted in ^{18}O relative to the hot brine (samples D, problem 2). Different still is steam from Growler Hot Spring near Mt. Lassen, which has δD and $\delta^{18}O$ values of −102 and −11.2 associated with hot chloride water that has values of −95 and −9.2, respectively (Janik et al., 1983). Use the fractionation factors in Table 3.2 to explain these data. What effects might salinity have on these results?

5. A 25°C air mass with a relative humidity of 70% contains water vapor with δD and $\delta^{18}O$ values of −107 and −15.0 per mil, respectively. Calculate the temperature of the dew point, and predict the δD and $\delta^{18}O$ values of the first dew that would form.

6. On a pleasant summer day, thermal convection produces fluffy cumulus clouds in an air mass that originally contained water vapor with δD and $\delta^{18}O$ values of −150 and −20.0 per mil, respectively. If the cloud has a temperature of 10°C and contains 0.9 g/m^3 of water droplets, what are the δD and $\delta^{18}O$ values of the droplets and the remaining vapor?

7. Use the information in Table 3.2 to determine the oxygen and hydrogen isotopic fractionation factors between ice and liquid water at 0°C.

8. A stirred vessel containing water with a δD value of -70 and a $\delta^{18}O$ value of -10.0 slowly undergoes freezing at 0°C. Using the fractionation factors in Table 3.2, calculate the δD and the $\delta^{18}O$ values of the water and the *average* ice when 50% of the water has been frozen. First, make this computation assuming that all the ice remains in isotopic equilibrium with the water. Second, redo this calculation assuming that the ice always forms in equilibrium with the well-stirred water, but that the progressive formation of new ice isolates any ice that was previously formed, thereby preventing any re-equilibration with the remaining reservoir of water.

9. An air mass that is undergoing Rayleigh condensation has lost 75% of its original water vapor, which initially had a δD value of -125. Assuming that the fractionation factor was very close to 1.100 throughout this entire process, what is the δD value of the water forming at this particular stage? If all the "rain" that was progressively removed from the system up to that time was collected in a single bucket, what average δD value would that water have?

10. Using Dansgaard's approximation (equation 3.33d), and the ice–vapor fractionation factors in Table 3.2, calculate the approximate δD and $\delta^{18}O$ values that would be observed at -10, -25, and -40°C from an atmospheric air mass that cooled isobarically from its dew point at 0°C at a total pressure of 0.8 atm. Use values of $-47.0‰$ and $-7.1‰$ for the initial condensate, to facilitate comparison of your results with the numerical calculations in Table 3.4. Determine the average slope of your calculated trend.

11. If the $\delta^{18}O$ variations of 1.65‰ in carbonates in deep-sea cores are presumed to be entirely due to the storage of glacial ice on continents, calculate the fraction of the ocean that was removed if the continental ice sheets had an average $\delta^{18}O$ value of (a) $-15‰$; (b) $-25‰$, and (c) $-35‰$. Which of these fractions, and which of the assumed $\delta^{18}O$ values for the ice sheets, are most compatible with other evidence? What if only half of the 1.65‰ fluctuations are attributable to ice storage and the rest to temperature changes in the world ocean?

12. Immediately prior to a storm that dropped 2 inches of rainfall with a $\delta^{18}O$ value of -9.7, a small stream that was flowing at 5 cfs had a $\delta^{18}O$ value of -5.0. A few hours later, when the stream peaked at 35 cfs, it had a $\delta^{18}O$ value of -6.5. Use simple arguments to estimate the relative contribution of "overland flow" contributed directly by the storm to the stream, based on (a) the volumetric stream-flow variations alone and (b) the isotope data alone. Explain the discrepancy.

13. Peak flow at a particular gauging station on a small river usually occurs about 32 h after intense rainfall events. Predict the time when streamflow will diminish to 35% of the peak flow. If you could make stable isotope and geochemical analyses of three samples to study the river, at what times would you collect them?

References

Aldaz, L. and Deutsch, S. (1967) On a relationship between air temperature and oxygen isotope ratio of snow and firn in the South Pole region. *Earth Planet. Sci. Lett.*, **3**, 267–274.

Barnola, J.M., Raynaud, D., Korotkevich, Y.S., and Lorius, C. (1987) Vostok ice core provides 160,000-year record of atmospheric CO_2. *Nature*, **329**, 408–414.

Bohlke, J.K. and Kistler, R.W. (1986) Rb-Sr, K-Ar, and stable-isotope evidence for the ages and sources of fluid components of gold-bearing quartz veins in the northern Sierra Nevada foothills metamorphic belt, California. *Econ. Geol.*, **81**, 296–322.

Bottinga, Y. and Craig, H. (1968) High temperature liquid–vapor fractionation factors for H_2O–HDO–H_2O^{18}. *Trans. Am. Geophys. Union*, **49**, 356–357.

Bowers, T.S. and Taylor, H.P., Jr. (1985) An integrated chemical and stable-isotope model of the origin of midocean ridge hot spring systems. *J. Geophys. Res.*, **90**, 12,583–12,606.

Clayton, R.N., Friedman, I., Graf, D.L., Mayeda, T.K., Meents, W.F., and Shimp, N.F. (1966) The origin of saline formation waters. I. Isotopic composition. *J. Geophys. Res.*, **71**, 3869–3882.

Clayton, R.N., O'Neil, J.R., and Mayeda, T.K. (1972) Oxygen isotopic exchange between quartz and water. *J. Geophys. Res.*, **77**, 3057–3067.

Craig, H. (1961) Isotopic variations in meteoric waters. *Science*, **133**, 1702–1703.

Craig, H. (1966) Isotopic composition and origin of the Red Sea and Salton Sea geothermal brines. *Science*, **154**, 1544–1548.

Craig, H. and Gordon, L.I. (1965) Deuterium and oxygen 18 variations in the ocean and the marine atmosphere. Symposium on Marine Geochemistry, Narraganset Marine Laboratory, University of Rhode Island Publication, Vol. 3, pp. 277–374.

Craig, H., Boato, G., and White, D.E. (1956) Isotopic geochemistry of thermal waters. In *Proceedings of the Conference on Nuclear Processes in Geological Settings.* 2nd National Academy of Science National Research Council Publication 400:29.

Criss, R.E. (1997) New formulation for the hydrograph, time constants for stream flow, and the variable character of "baseflow." *Trans. Am. Geophys. Union*, **78**, 317.

Criss, R.E. and Davisson, M.L. (1996) Isotopic imaging of surface water/groundwater interactions, Sacramento Valley, California. *J. Hydrol.*, **178/1–4**, 205–222.

Dansgaard, W. (1964) Stable isotopes in precipitation. *Tellus*, **16**, 436–468.

Dansgaard, W., Johnsen, S.J., Miller, J., and Langway, C.C. (1969) One thousand centuries of climatic record from Camp Century on the Greenland ice sheet. *Science*, **166**, 377–381.

Davisson, M.L. and Criss, R.E. (1996) Na–Ca–Cl relations in basinal fluids. *Geochim. Cosmochim. Acta*, **60**, 2743–2752.

Davisson, M.L., Presser, T.S., and Criss, R.E. (1994) Geochemistry of tectonically expelled fluids from the northern Coast Ranges, Rumsey Hills, California. *Geochim. Cosmochim. Acta*, **58**, 1687–1699.

Emiliani, C. and Shackleton, N.J. (1974) The Bruhnes epoch: isotopic paleotemperatures and geochemistry. *Science*, **183**, 511–514.

Epstein, S. and Mayeda, T.K. (1953) Variations of O^{18} content of waters from natural sources. *Geochim. Cosmochim. Acta*, **4**, 213–224.

Epstein, S., Sharp, R.P., and Goddard, I. (1963) Oxygen isotope ratios in Antarctic snow, firn, and ice. *J. Geol.*, **71**, 698–720.

Epstein, S., Sharp, R.P., and Gow, A.J. (1965) Six-year record of oxygen and hydrogen isotope variations in South Pole firn. *J. Geophys. Res.*, **70**, 1809–1814.

Epstein, S., Sharp, R.P., and Gow, A.J. (1970) Antarctic ice sheet: stable isotope analysis of Byrd Station cores and interhemispheric climatic implications. *Science*, **168**, 1570–1572.

Ferronsky, V. I. and Polyakov, V.A. (1982), *Environmental Isotopes in the Hydrosphere*. John Wiley & Sons, New York.

Frederickson, G.C. and Criss, R.E. (1999) Isotope hydrology and time constants of the unimpounded Meramec River basin, Missouri. *Chem. Geol.* (in press).

Gat, J.R. and Carmi, I. (1970) Evolution of the isotopic composition of atmospheric waters in the Mediterranean Sea area. *J. Geophys. Res.*, **75**, 3039–3048.

Gregory, R.T. and Taylor, H.P., Jr. (1981) An oxygen isotope profile in a section of Cretaceous oceanic crust, Samail ophiolite, Oman: evidence for $\delta^{18}O$-buffering of the oceans by deep (> 5 km) seawater-hydrothermal circulation at mid-ocean ridges. *J. Geophys. Res.*, **86**, 2737–2755.

Hedenquist, J.W., Aoki, M., and Shinohara, H. (1994) Flux of volatiles and ore-forming metals from the magmatic-hydrothermal system of Satsuma Iwojima volcano. *Geology*, **22**, 585–588.

Ingraham, N.L. and Taylor, B.E. (1991) Light stable isotope systematics of large-scale hydrologic regimes in California and Nevada. *Water Resour. Res.*, **27**, 77–90.

Janik, C.J., Nehring, N.L., and Truesdell, A.H. (1983) Stable isotope geochemistry of thermal fluids from Lassen Volcanic National Park, California. *Geothermal Resour. Council, Trans.*, **7**, 295–300.

Johnsen, S.J., Dansgaard, W., Clausen, H.B., and Langway, C.C. (1972) Oxygen isotope profiles through the Antarctic and Greenland ice sheets. *Nature*, **235**, 429–434.

Joseph, A., Frangi, J.P., and Aranyossy, J.F. (1992) Isotope characteristics of meteoric water and groundwater in the Sahelo-Sudanese zone. *J. Geophys. Res.*, **97**, 7543–7551.

Jouzel, J., Lorius, C., Petit, J.R., Genthon, C., Barkov, N.I., Kotlyakov, V.M., and Petrov, V.M. (1987) Vostok ice core: a continuous isotope temperature record over the last climatic cycle (160,000 years). *Nature*, **329**, 403–408.

Kamilli, R.J. and Criss, R.E. (1996) Genesis of the Silsilah tin deposit, Kingdom of Saudi Arabia. *Econ. Geol.*, **91**, 1414–1434.

Kelley, W.C., Rye, R.O., and Livnat, A. (1986) Saline minewaters of the Keweenaw Peninsula, northern Michigan: their nature, origin, and relation to similar deep waters in Precambrian crystalline rocks of the Canadian shield. *Am. J. Sci.*, **286**, 281–308.

Kharaka, Y.K. and Carothers, W.W. (1986) Oxygen and hydrogen isotope geochemistry of deep basin brines. In *Handbook of Environmental Isotope Geochemistry*, Vol. 2, Fritz, P. and Fontes, J.Ch., eds., Elsevier, New York, pp. 305–360.

Lawrence, J.R. and Gieskes, J.M. (1981) Constraints on water transport and alteration in the oceanic crust from the isotopic composition of pore water. *J. Geophys. Res.*, **86**, 7924–7934.

Lorius, C., Jouzel, J., Ritz, C., Merlivat, L., Barkov, N.I., Korotkevich, Y.S., and Kotlyakov, V.M. (1985) A 150,000-year climatic record from Antarctic ice. *Nature*, **316**, 591–596.

Majoube, M. (1970) Fractionation factor of ^{18}O between water vapour and ice. *Nature*, **226**, 1242.

Majoube, M. (1971) Fractionnement en oxygene 18 et en deuterium entre l'eau et sa vapeur. *J. Chim. Phys.*, **68**, 1423–1436.

Melchiorre, E.B., Criss, R.E., and Davisson, M.L. (1999) Relationship between seismicity and subsurface fluids, central California Coast Ranges. *J. Geophys. Res.*, **104**, 921–939.

Merlivat, L. and Nief, G. (1967) Fractionnement isotopique lors des changements d'état solide–vapeur et liquide–vapeur de l'eau à des temperatures inferieures à 0°C. *Tellus*, **19**, 122–126.

Muehlenbachs, K. and Clayton, R.N. (1976) Oxygen isotope composition of the oceanic crust and its bearing on seawater. *J. Geophys. Res.*, **81**, 4365–4369.

O'Neil, J.R. and Taylor, H.P., Jr. (1969) Oxygen isotope equilibrium between muscovite and water. *J. Geophys. Res.*, **74**, 6012–6022.

O'Neil, J.R., Clayton, R.N., and Mayeda, T.K. (1969) Oxygen isotope fractionation in divalent metal carbonates. *J. Chim. Phys.*, **51**, 5547–5558.

Petit, J.R., Basile, I., Leruyuet, A., Raynaud, D., Lorius, C., Jouzel, J., Stievenard, M., Lipenkov, V., Barkov, N., Kudryashov, B., Davis, M., Saltzman, E., and Kotlyakov, V. (1997) Four climate cycles in Vostok ice core. *Nature*, **387**, 359–360.

Picciotto, E., DeMaere, X., and Friedman, I. (1960) Isotopic composition and temperature of formation of Antarctic snow. *Nature*, **187**, 857–859.

Rayleigh, Lord (1902) On the distillation of binary mixtures. *Phil. Mag.*, S.6, v. 4, 521–537.

Rose, T.P. and Davisson, M.L. (1996) Radiocarbon in hydrologic systems containing dissolved magmatic carbon dioxide. *Science*, **273**, 1367–1370.

Rose, T.P., Davisson, M.L., and Criss, R.E. (1996) Isotope hydrology of voluminous cold springs in fractured rock from an active volcanic region, northeastern California. *J. Hydrol.*, **179**, 207–236.

Savin, S.M. and Yeh, H.W. (1981) Stable isotopes in ocean sediments. In *The Sea*, Vol. 7, Emiliani, C., ed. John Wiley & Sons, New York, pp. 1521–1554.

Shackleton, N.J. and Pisias, N.G. (1985) Atmospheric carbon dioxide, orbital forcing, and climate. *Am. Geophys. Union, Geophys. Mon.*, **3**, 303–317.

Sheppard, S.M.F. (1986) Characterization and isotopic variations in natural waters. *Rev. Mineral.*, **16**, 165–183.

Sheppard, S.M.F., Nielsen, R.L., and Taylor, H.P., Jr. (1971) Oxygen and hydrogen isotope ratios in minerals from porphyry copper deposits. *Econ. Geol.*, **66**, 515–542.

Sklash, M.G. and Farvolder, R.N. (1979) The role of groundwater in storm runoff. *J. Hydrol.*, **43**, 45–65.

Taylor, B.E. (1986) Magmatic volatiles: isotopic variations of C, H and S. *Rev. Mineral.*, **16**, 185–219.

White, D.E., Muffler, L.J.P., and Truesdell, A.H. (1971) Vapor dominated hydrothermal systems compared with hot water systems. *Econ. Geol.*, **66**, 75–97.

4

Nonequilibrium Fractionation and Isotopic Transport

4.1 Kinetics of Isotopic Exchange

At the Earth's surface, isotopic disequilibrium is far more common than isotopic equilibrium. Although isotopic equilibrium is approached in certain instances, numerous constituents of the lithosphere, hydrosphere, atmosphere, and biosphere are simply not in mutual isotopic equilibrium. This condition is consistent with the complex and dynamic conditions typical of the Earth's surface, particularly the large material fluxes, the rapid changes in temperature, and the biological mediation of chemical systems.

Fortunately, several aspects of isotopic disequilibrium may be understood in terms of elementary physical laws. For homogeneous phases such as gases or well-stirred liquids, or for cases where spatial gradients in isotopic contents are not of primary interest, then the principles of elementary kinetics can be applied. For cases where isotopic gradients are important, the laws of diffusion are applicable.

4.1.1 Isotopic Exchange Law

If two phases are out of isotopic equilibrium, they will progressively tend to approach the equilibrium state with the passage of time. This phenomenon occurs by the process of isotopic exchange, and its rate may be understood by examining isotopic exchange reactions from the viewpoint of elementary kinetic theory. In particular, consider the generalized exchange reaction

$$A + B^* \underset{k}{\overset{k\alpha}{\rightleftarrows}} A^* + B \qquad (4.1)$$

where A and B are two phases that share a common major element, and A^* and B^* represent the same phases in which the trace isotope of that element is present. The present analysis is simplified if the exchange reaction is written so that only one atom is exchanged, in which case the stoichiometric coefficients are all unity.

For reaction 4.1, kinetic principles assert that the forward and reverse reactions do not, in general, proceed at identical rates, but rather at the rates indicated by the quantities $k\alpha$ and k written by the arrows, multiplied by the appropriate concentrations terms. Assuming that the reaction is first order, then the reaction progress, represented by the quantity dA^*/dt, may be expressed by the difference between these forward and reverse rates, as follows:

$$dA^*/dt = k\alpha(A)(B^*) - k(A^*)(B) \qquad (4.2)$$

In order to evaluate the exchange process more completely, is important to carefully chose a consistent set of concentrations for substitution equation 4.2. Most workers have used extensive variables such as "moles" for this purpose, which can lead to an unrealistic dependence of exchange rates on the system size. Such problems are eliminated by the use of intensive variables. In particular, Criss et al. (1987) have shown that a realistic exchange law may be developed using the thermodynamic activities of the relevant isotopic species. For an appropriately written exchange reaction, the ratios of the activities are

$$R_A = A^*/A \qquad (4.3a)$$

and

$$R_B = B^*/B \qquad (4.3b)$$

where R_A and R_B represent the isotope ratios. If these ratios are introduced into the activity product of equation 4.1, the equilibrium condition is immediately obtained:

$$\alpha_{A-B} = \frac{[A^*][B]}{[A][B^*]} = \frac{R_A}{R_B} \qquad (4.4)$$

Equation 4.2 also reduces to this latter condition when dA^*/dt is zero, conforming to the required condition that the net reaction rate is zero when equilibrium is attained.

The exchange law (equation 4.2) may be expressed in terms of isotope ratios by differentiating equation 4.3a:

$$dR_A = \frac{dA^*}{A} - \frac{A^* \, dA}{A^2} \qquad (4.5a)$$

For situations where A^* is a trace isotope that has a minuscule concentration, then the second term on the right-hand side of equation 4.5a is negligible, so that

$$dR_A = \frac{dA^*}{A} \qquad (4.5b)$$

Finally, substituting the relationships given in equations 4.3a and b and 4.5b into equation 4.2 gives

$$dR_A/dt = -kB(R_A - \alpha_{A-B}R_B)$$

(4.6)

Equation 4.6 is the generalized differential equation that governs isotopic exchange. According to this equation, at any particular instant, the rate of change of R_A is directly proportional to the difference between the instantaneous isotopic ratio of phase A and the value that would be in equilibrium with phase B at that time. The proportionality constant between these quantities, kB, represents the product of the rate constant k and the activity B of the substance with which the phase of interest is exchanging. As discussed below, for cases where condensed phases are involved, k commonly contains a hidden term for the surface area, or the surface area-to-volume ratio.

An important detail is that equation 4.6 is the correct law of isotopic exchange, even for cases where the actual isotope ratios differ from the ratios A^*/A and B^*/B because of symmetry numbers that do not cancel. In such a case, α in both equations 4.1 and 4.4 has to be replaced by the equilibrium constant K, but, according to the Ferronsky and Polyakov rule, this K would differ from α by these same symmetry numbers. It is easy to show that equation 4.6 is also the correct mathematical result for such cases.

As a result, equation 4.6 is applicable to a large number of isotopic systems, including those where the isotopic contents of both phases are simultaneously changing. This equation may be modified and solved for many different problems, as developed in the following.

4.1.2 Condensed Phases

Pure solids and liquids have unit activities. For problems that involve exchange of a trace isotope among such substances, then the activities of the common isotopic form of the solids or liquids will still be essentially unity. A consistent set of activities for the relevant species is therefore the following:

$$A = 1 \qquad B = 1 \qquad A^* = R_A \qquad B^* = R_B$$

(4.7)

where R_A and R_B represent the isotope ratios. Note that the relationships in equation 4.3 and 4.4 are still satisfied. For this situation, equation 4.6 simplifies to the rate law of Criss et al. (1987):

$$dR_A/dt = -k(R_A - \alpha R_B)$$

(4.8)

This equation is very useful for fluid–mineral exchange. Two important solutions to this equation are given in the following sections; additional examples are given in chapter 5 and in Gregory et al. (1989).

4.1.2.1 Case 1: Infinite Reservoir   Equation 4.8 may be integrated directly in the case where R_B is constant. Such a situation could occur if the isotope ratio of phase B is buffered or if an infinite reservoir of phase B is present. In this case, the quantity αR_B is also constant and represents the value R_A^{eq} that phase A will attain at equilibrium (i.e., after infinite time). For this case, the solution to equation 4.8 is, for constant k,

$$\frac{R_A - R_A^{eq}}{R_A^i - R_A^{eq}} = e^{-kt} \tag{4.9a}$$

where R_A^i is the initial value at $t = 0$. The ratio on the left-hand side of equation 4.9a is also identical to the analogous ratio of the δ-values, provided that the appropriate subscripts are all retained (equation A.2). That is,

$$\frac{\delta_A - \delta_A^{eq}}{\delta_A^i - \delta_A^{eq}} = e^{-kt} \tag{4.9b}$$

The left-hand side is also equal to the quantity $1 - F$, where F is a progress variable that represents the fractional approach to equilibrium. Note that $F = 0$ at the beginning of the exchange process and that $F = 1$ at equilibrium (very long times). Thus,

$$\ln(1 - F) = -kt \tag{4.9c}$$

4.1.2.2 Case 2: Closed System   In the process of isotopic exchange, the gain of a heavy isotope by one phase must occur at the expense of the heavy isotope content of another. In the case of a simple closed system, the total amount of the heavy isotope in the system is constant. For a binary system, material balance in a closed system requires that

$$X_A \, dR_A = -X_B \, dR_B \tag{4.10}$$

where X_A and X_B are the mole fractions of the element of interest for phases A and B, respectively. This simple relationship may be obtained by differentiating the material balance equation (equation 1.26a), recognizing that the mole fractions and the isotopic ratio of the overall system are constants. The simultaneous solution to differential equations 4.8 and 4.10 is (Criss et al., 1987)

$$\frac{R_A - R_A^{eq}}{R_A^i - R_A^{eq}} = e^{-kt(\alpha X_A + X_B)/X_B} \tag{4.11a}$$

where k has again been assumed to be constant. Note that the left-hand side of this equation is again identical to $1 - F$. Finally, if the fractionation factor α is close to unity, and because in a two-phase system the sum of X_A and X_B is unity, this relationship is very nearly equal to

$$\frac{R_A - R_A^{eq}}{R_A^i - R_A^{eq}} = e^{-kt/X_B} \tag{4.11b}$$

This result reduces to equation 4.9a for an infinite reservoir because $X_B \rightarrow 1.0$.

4.1.3 Gas–Solid and Gas–Liquid Exchange

Many processes of interest involve isotopic exchange of a trace isotope between a gas and a solid or liquid phase. For such systems, the activities of the gaseous species will equal the fugacities of the relevant isotopomers, f^* and f. In contrast, the ordinary form of the solid or liquid phase will have a unit activity, while the species of this phase that bears the trace isotope will have the activity R_B. Thus, the relevant activities of the participating species are

$$A = f_A \qquad B = 1 \qquad A^* = f_A^* \qquad B^* = R_B \qquad (4.12a)$$

A reasonable assumption is that the isotopic ratio R_A is either equivalent or directly proportional to the ratio of the fugacity of the gas species that has the trace isotope to that of the common gas; any effects that involve symmetry numbers will ultimately cancel. Moreover, because the fugacity coefficients will cancel because of the ideality of isotopic mixtures, this ratio will be identical to the ratio of the partial pressures, here denoted by P^* and P, of the relevant gaseous species. Thus, the isotope ratio R_A of the gas is

$$R_A = A^*/A = f_A^*/f_A = P_A^*/P_A \qquad (4.12b)$$

while that for the condensed phase is, again,

$$R_B = B^*/B \qquad (4.12c)$$

In this case, the law that describes the rate of change of the gas will follow differential equation 4.8, as will be demonstrated in the following sections on water–water vapor and water–carbon dioxide exchange. For cases where a small amount of gas is equilibrating with a large reservoir of a solid or liquid, the appropriate solution to this differential equation is equation 4.9.

An even more interesting situation arises if the rate of change of the condensed phase is measured, so that phase A must now represent the condensed phase. In this case, equation 4.6 must be used, and the relevant activities are

$$A = 1 \qquad B = f_B \qquad A^* = R_A \qquad B^* = f_B^* \qquad (4.12d)$$

This seemingly subtle switch made between equations 4.12a and 4.12d has an important ramification, for in this case the value of B is not unity, but rather equals the partial pressure of the gas. Equation 4.6 therefore predicts a direct proportionality between the exchange rate of the condensed phase and this partial pressure f_B, which has been experimentally confirmed (see later).

4.1.4 Exchange among Gaseous Phases

For problems that involve isotopic exchange among gaseous substances, it is likely that the fugacities of the relevant species should be used in every case. The equivalence or direct proportionality between the fugacity ratios A^*/A and B^*/B and the isotope ratios, R_A and R_B would still be retained. The appropriate activities are therefore

$$A = f_A \qquad B = f_B \qquad A^* = f_A^* \qquad B^* = f^*_B \qquad (4.13a)$$

In addition,

$$R_A = A^*/A = f_A^*/f_A = P_A^*/P_A \qquad (4.13b)$$

and

$$R_B = B^*/B = f_B^*/f_B = P_B^*/P_B \qquad (4.13c)$$

It is probable that the rate law will follow equation 4.6, but this requires experimental confirmation. Gaseous systems are notorious for not progressing toward isotopic equilibrium at appreciable rates. This situation arises because, at low to moderate temperatures, there are few mechanisms that promote isotopic exchange among the independent molecules. Seemingly, either formation of an intermediate complex or disassociation of the molecules would be required for isotopic exchange among the gases to occur, and such processes are simply not likely to occur in low-energy environments. As a result, the coexisting methane, hydrogen, and hydrogen sulfide in gas wells are not close to hydrogen isotope equilibrium; the oxygen, carbon dioxide, and ozone in ordinary air are not close to oxygen isotope equilibrium, and so on. Of course, it is possible that a catalyst could be used to promote such exchange reactions.

4.1.5 Large Concentrations of Minor Isotope

If the minor isotope is present in a significant concentration, as is the case for the stable isotopes of chlorine, then the differential equation must be modified. In this case, both terms on the right-hand side of equation 4.5a must be retained. because A^* is not a trace isotope. However, the result may be simplified because the sum of A plus A^* must be a constant for an exchange reaction, and, consequently,

$$dA^* = -dA \qquad (4.14a)$$

Introducing the last result into equation 4.5a gives

$$dR_A = \frac{dA^*}{A}(1 + R_A) \qquad (4.14b)$$

Substituting this result into equation 4.2 gives (R.E.C., unpublished)

$$\boxed{\frac{d\ln(1 + R_A)}{dt} = -(kB)(R_A - \alpha_{A-B}R_B)} \qquad (4.15)$$

Equation 4.15 is the hypothetical differential equation that governs isotopic exchange in systems where the minor isotope is not present in trace quantities. The result is similar to equations 4.6 and 4.8, in that the exchange rate is directly proportional to the difference between the actual isotopic ratio of phase A and its equilibrium value at that instant, but in this case it is $d\ln(1 + R)$, rather than $d\ln R$, that is proportional to this factor. Note that this equation correctly reduces to equation 4.6 for a trace isotope, because $\ln(1 + R) \approx R$ for small R. As such, equation 4.15 provides a more general and exact expression of the rate of isotopic exchange.

Depending on the system, the activity of B might be unity; in other cases, the activity of B might play a critical role in governing the exchange process. This equation requires experimental verification for an element, such as chlorine, that has an abundant minor isotope.

4.1.6 Dependence of Rate Constants on Physical Parameters

4.1.6.1 Temperature Dependence   Isotopic exchange phenomena generally depend on temperature, as can be seen from equation 4.6. One effect is that the isotopic fractionation factor, which is included in the rate equations, depends on temperature. Another effect is that the chemical activity of "B" in equation 4.6 may be temperature dependent. Finally, the rate constant k may be temperature dependent. For numerous situations, the variation of the rate constant with temperature follows an Arrhenius law:

$$k = A_0 e^{-E_a/RT}$$

(4.16)

where A_0 is a constant, E_a is the activation energy, R is the gas constant, and temperature T is in kelvins. Diffusion constants, which in many cases are directly proportional to k, commonly also follow this type of temperature dependence. Several examples of such effects are compiled in Cole and Ohmoto (1986).

4.1.6.2 Pressure Dependence   Isotopic exchange rates can depend strongly on pressure. For example, the relative thermodynamic activities of different gases in mixtures may be pressure dependent, which can modify the value of "B" in equation 4.6. More important, large increases in exchange rates between several silicate minerals and water have been observed at high pressures. These effects arguably arise because the rates of diffusion of molecular water or OH^- are significantly enhanced at high pressures, as reviewed by Cole and Ohmoto (1986). Of course, in geologic environments where materials are subjected to mechanical deformation, exchange rates may be greatly enhanced.

4.1.6.3 Salt Concentrations   Isotopic exchange rates in systems that involve water can depend in a complex manner on the concentration of dissolved salts. First, the concentration of salts can change the activity of water, which in many cases will cause "B" in equation 4.6 to have a value different than unity. Similarly, a change in the activity of water will generally lower the saturation vapor pressure of water above the solution, generally in direct proportion to the mole fraction of the water. Such effects, to first order, depend only on the mole fractions of salt and water, and do not strongly depend on the nature of the salt.

In addition, the presence of salts can significantly change equilibrium fractionation factors. For example, the fractionation factor between an aqueous solution and carbon dioxide will differ from the factor between pure water and carbon dioxide. Such effects are variable and depend on the concentration, as well as the type of salt. Many references are compiled in Friedman and O'Neil (1977).

In certain instances, salt concentrations can have even more profound effects on exchange rates. The oxygen isotope exchange rate between alkali feldspar and water can be greatly promoted by placing the feldspar in a solution that is far from cation equilibrium with the mineral (O'Neil and Taylor, 1967). In contrast, the presence of calcium chloride greatly retards the rate of equilibration of carbon dioxide with water (see below).

4.2 Examples of Isotopic Exchange Kinetics

4.2.1 Carbon Dioxide–Water Exchange

The rate of oxygen isotope exchange between carbon dioxide gas and water can be studied with equipment that is standard in many stable isotope laboratories. This exchange system rapidly attains equilibrium, and provides the routine means of analyzing oxygen isotopes in water samples (Epstein and Mayeda, 1953).

Figure 4.1 shows results for CO_2-water exchange at 25°C, measured as a function of time from the initial juxtaposition of the phases. The raw $\delta^{18}O$ value of the starting CO_2 gas was -0.4, while that of the water was -30.5. The latter value will not be measurably affected by the exchange, because water is present in great excess in this system. However, over time, the $\delta^{18}O$ value of the CO_2 gas progressively increases until it reaches its final equilibrium value of $+9.53$. Note that the

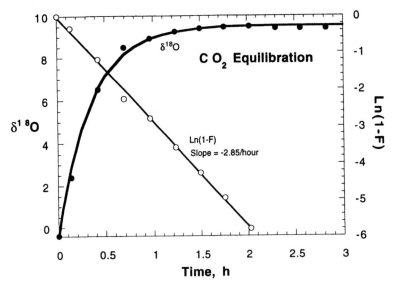

Figure 4.1 Exchange of pure gaseous CO_2 with liquid water at 25°C. The mole fraction of water dominates in this 25-ml closed vessel, charged with 1 atm of CO_2 and with 5 ml of water. Over time, the $\delta^{18}O$ value of the CO_2 changes from its initial value of -0.40 to $+9.53‰$, with the latter representing equilibrium with the water. The curve was calculated with equation 4.9b, given the empirical value of 2.85/h that is indicated for the rate constant k (equation 4.9c, right scale).

equilibrium fractionation factor between carbon dioxide gas and water is 1.0412 at 25°C.

The curve in Figure 4.1 represents equation 4.9b, calculated for an empirical value of k of 2.85/hr. It is evident that the progress toward the equilibrium condition is exponential. The rate constant is likely proportional to the extent of the contact between the water and the CO_2 gas, which increases over time, and which can be augmented by increasing the surface area or by shaking or stirring the water.

Fortier (1994) has shown that the rate of isotopic exchange of the CO_2 does not depend strongly on the partial pressure of CO_2 gas. This result is in accord with the previous results, in that it has been shown that equation 4.6 reduces to equation 4.8 for this case. Fortier also demonstrated that the rate of exchange depends on the agitation rate of the water, and, given rapid agitation rates, on the temperature and on the concentration of dissolved $CaCl_2$ salt. For agitated samples, Fortier's result for his particular experimental apparatus fit the empirical relationship

$$\ln k = 0.52 - 0.63\,m - 2100/T \tag{4.17}$$

where k is in inverse seconds, the molality m is in mol/kg, and T is in kelvins. The temperature dependence is appropriate for an Arrhenius relationship. The decrease in exchange rate with concentration is likely due to the greater difficulty in achieving exchange with water molecules that are solvated with dissolved ions (Fortier, 1994). Interestingly, this latter decrease is much greater than can be accounted by equation 4.6 for a simple decrease in the activity of water in the solutions.

4.2.2 Water–Water Vapor Exchange

Isotopic exchange between water and water vapor is attained on short timescales. Equations 4.9a and b describe the situation where a small amount of vapor at 100% humidity with an isotope ratio R_v is equilibrating with a large liquid water reservoir with ratio R_w. In this case, the isotope ratio of the equilibrium vapor, R_v^{eq}, is simply equal to Rw/α.

The case where a small body of water equilibrates with a large, saturated vapor reservoir is more interesting. In this case, the appropriate differential equation developed from equations 4.6 and 4.12d is

$$\frac{dR_w}{dt} = -k'(R_w - \alpha R_v) \tag{4.18a}$$

where:

$$k' = k\left(\frac{s_w}{v_w}\right)p^0 \tag{4.18b}$$

where R_w and R_v, respectively, refer to the isotope ratios of the water and vapor, and αR_v represents the value the water would have were it in equilibrium with the vapor at that instant. Moreover, the overall rate constant k' incorporates several factors, including the surface area-to-volume ratio of the water body, as well as the fugacity—here taken to be the saturation vapor pressure of the water at the temperature of interest. The solution is simply

$$\frac{R_w - R_w^{eq}}{R_w^i - R_w^{eq}} = e^{-k't} \tag{4.19}$$

where k' is again given in equation 4.18b. Ingraham and Criss (1993, 1998) have shown that this expression quantitatively explains experimental data. They also demonstrated that, for a given experiment, the value of k is approximately 0.0086 cm/torr-day at 20°C, and that this value is identical for D/H, $^{18}O/^{16}O$, and T/H exchange.

Water–water vapor exchange effects are elucidated by an experimental setup where two isotopically distinct waters are sealed in a closed system (Figure 4.2; Ingraham and Criss, 1993). The humidity in the box quickly approaches 100% and the vapor achieves a composition intermediate between the two equilibrium vapors appropriate for the two beakers. Because neither water is in isotopic equilibrium with this mean vapor, both beakers change in composition by exchange. In effect, the waters exchange with each other, by the intermediary vapor phase, until they both ultimately attain a composition equal to their volu-metrically weighted mean composition (Figure 4.3). Increasing the temperature of the box significantly increases the exchange rate, in direct proportion to the magnitude of the saturation vapor pressure, in accordance with the behavior predicted by equation 4.18b.

Other experiments conducted by Ingraham and Criss (1993) demonstrate the exchange of beakers of isotopically distinct waters that have differing surface areas or volumes and that were placed in a sealed box (Figure 4.4). In this case, for each water body the differential equation is

$$\frac{dR_w}{dt} = -k(R_w - R_M) \tag{4.20a}$$

R a

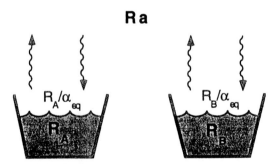

Figure 4.2 Isotopic exchange between two water samples that have different isotope ratios R_A and R_B is realized by the exchange of each water with the ambient vapor phase in a closed system. The isotope ratio of the vapor, R_a, differs from the two equilibrium values R_a/α and R_B/α that develop in a thin boundary layer immediately above the water surfaces. Over time, the waters progress toward a mean isotopic composition that is in equilibrium with the vapor phase. After Ingraham and Criss (1993).

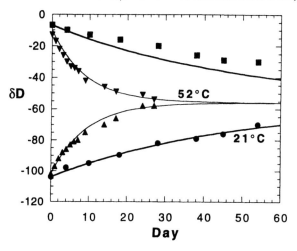

Figure 4.3 Graph of δD values versus time for two water beakers that have equal surface areas and equal volumes and that are undergoing isotopic exchange in sealed systems. In both experiments, the waters did not undergo significant evaporation, but their isotopic ratios progress toward their average value of $-56‰$ via the process of exchange with ambient atmosphere vapor. The rate constant for the 52°C experiment was 5.5× greater than that for the 21°C exchange experiment, conforming to the ratio of the vapor pressure of water at the two temperatures (102.08:18.65 torr), as predicted by factor "B" in equation 4.6. Solid curves were calculated by equations 4.18b and 4.19. Experimental data (points) are from Ingraham and Criss (1998).

where R_w is the instantaneous isotope ratio of the water, and R_M is the instantaneous equilibrium isotope ratio R_M that is the same for all the waters in the system. However, the value of R_M will generally change during the course of the exchange experiment, because it is a function of both the surface areas and the isotope ratios of the various waters, according to

$$R_M = \frac{\Sigma s_j R_j}{\Sigma s_j} \qquad (4.20\text{b})$$

where the sums are computed over all the distinct water bodies. In a small closed system, over time, the waters will approach a common equilibrium value, which is the volumetrically weighted average isotope ratio of the water bodies initially placed in the system:

$$R_{eq} = \frac{\Sigma v_j R_j^i}{\Sigma v_j} \qquad (4.20\text{c})$$

A mathematical solution to equations 4.20a, b, and c was found and experimentally confirmed by Ingraham and Criss (1993) for the case of two water bodies in a closed system. For this case,

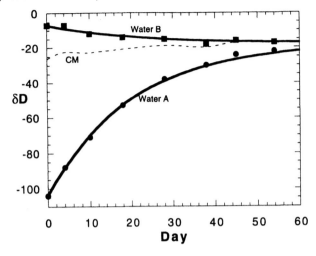

Figure 4.4 Graph of δD values versus time for two water beakers that have equal surface areas but unequal volumes (~5:1) and that are undergoing isotopic exchange in a sealed system. The isotopic ratios progress toward their volumetrically weighted average value over time via exchange with ambient atmosphere vapor. This produces an asymmetrical pattern because the smaller sample (water A) must undergo the greatest isotopic shift (cf. Figure 4.3). Because the system was not perfectly sealed, the waters underwent some evaporation, producing small variations in the δD value of the system (−22 ± 4; dotted curve) over time. Solid curves were calculated by equations 4.20 and 4.21. Experimental data (points) are from Ingraham and Criss (1993)

$$\frac{R_w - R_w^{eq}}{R_w^i - R_w^{eq}} = e^{-kGP^0 t} \qquad (4.21a)$$

where the parameter G is

$$G = \left(\frac{s_A s_B}{s_A + s_B}\right)\left(\frac{v_A + v_B}{v_A v_B}\right) \qquad (4.21b)$$

Note that G has the units of inverse length, as does the surface area-to-volume ratio. However, in this case, the surface area term and the volume term each assume a form that is mathematically analogous to the reduced mass. An experimental example is shown in Figure 4.4.

4.3 Evaporation

The isotopic exchange of water with saturated water vapor is one endmember process of the more complex phenomenon of evaporation. Previous workers have successfully studied this process from the viewpoint of diffusion theory. Elementary kinetic theory can provide results of comparable value, because isotopic gradients are normally not measured in such systems. Moreover, the kinetic theory is simple to develop, involves fewer assumptions, and logically builds upon the arguments established earlier in this chapter.

Evaporation may be viewed in terms of the rates of transport of water molecules from the water to the vapor phase, and by reverse transport of molecules from the vapor back to the water (Figure 4.5). Where chemical equilibrium is attained, so that the humidity h is unity (i.e., 100%), there is no net evaporation. In this case, the forward and reverse rates are not zero but rather are equal in magnitude; this very process promotes isotopic exchange between liquid water and vapor, as described previously. However, when the humidity is less than unity, the return rate becomes lower than the loss rate, and net evaporation occurs.

This verbal description may be translated into a plausible physical theory, with the result applicable to both hydrogen and oxygen isotopes (R. E. C., unpublished, 1975). The net rate of loss of ordinary $H_2{}^{16}O$ from the water body, here denoted by dN/dt, is given by the difference between these forward (loss) and reverse (return) rates. The forward rate will be proportional to the activity of $H_2{}^{16}O$ in the water, W, which is unity. The return rate will be proportional to the partial pressure P_v of $H_2{}^{16}O$ molecules in the vapor phase. Thus,

$$-dN/dt = k_1 P_v - k_2 W \tag{4.22a}$$

where k_1 and k_2 are rate constants. Similar arguments may be made for the trace isotopic species of water, whether it be $H_2{}^{18}O$ or HDO, as follows:

$$-dN^*/dt = k_3 P_v^* - k_4 W^* \tag{4.22b}$$

where k_3 and k_4 are additional rate constants. For this situation, the following relationships apply:

$$R_w = W^*/W \qquad R_v = P_v^*/P_v \tag{4.23}$$

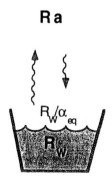

Figure 4.5 Diagram that shows isotopic effects during evaporation of water. The isotope ratios for the water (R_w), atmosphere water vapor (R_a), and vapor (R_w/α) in the thin, saturated boundary layer immediately above the water surface are indicated. If the humidity is less than 100%, the transport rate of water molecules away from the water body is greater than the rate of condensation of vapor onto the water surface, as schematically indicated by the wavy arrows, and net evaporation occurs.

The problem reduces to manipulation of these phenomenological equations and evaluation of the four constants. This may be readily accomplished by examination of the relationships between the water and water vapor that occur (1) at chemical and isotopic equilibrium, where the humidity h is unity; and (2) at zero humidity. At equilibrium, there is no net loss of liquid water, so dN/dt and dN^*/dt are both zero. As a result, equations 4.22a and b, respectively, become

$$k_1 P_v = k_2 W \tag{4.24a}$$

and

$$k_3 P_v^* = k_4 W^* \tag{4.24b}$$

where P_v and P_v^* refer to the vapor pressures of ordinary water and the isotopomer. Examined in the two limiting cases where the two water samples are isotopically pure; that is, where W and W^* are each equal to unity, these equations become

$$P_v^0 = k_2/k_1 \tag{4.25a}$$
$$P_v^{*0} = k_4/k_3 \tag{4.25b}$$

where P_v^0 and P_v^{*0} refer to the vapor pressures above the pure liquids. Now, dividing equation 4.24b by equation 4.24a, or, alternatively, dividing equation 4.25b by equation 4.25a and using equations 3.8 or 3.12, establishes an important relationship among the k values:

$$\alpha_{eq} = \frac{k_2 k_3}{k_1 k_4} \tag{4.26}$$

Now, consider the situation of evaporation of water into an atmosphere of zero humidity, where P_v^* and P_v are both zero, so that there is no return transport of atmospheric water vapor to the water. Making these substitutions and then dividing equation 4.22b by equation 4.22a gives

$$dN^*/dN = k_4 R_w/k_2 \tag{4.27a}$$

The ratio dN^*/dN is the quantity R_E, and represents the isotopic ratio of the water that is lost to the dry atmosphere at any instant. Under these dry conditions, the quotient R_w/R_E is the nonequilibrium fractionation factor, hereafter denoted by α_{evap}^0, for the system. Craig et al. (1963) have measured this factor by evaporating water into dry nitrogen gas, and analyzing the composition of the remaining water as a function of time (see below). In this case, the isotopic ratio of the water follows a Rayleigh curve. Here, the quantity f represents the fraction of water that remains to be evaporated, while the exponent of f is the quantity $(1/\alpha_{evap}^0 - 1)$, as described in equation 4.28. Consequently, equation 4.27a reduces to the following condition:

$$\alpha_{evap}^0 = (R_w/R_E)_{h=0} = k_2/k_4 \tag{4.27b}$$

Combining this result with equation 4.26 gives another important relationship among the k values:

$$\alpha_{evap}^0 = \alpha_{eq}(k_1/k_3) = k_2/k_4 \tag{4.27c}$$

Now that these relationships have been established among the constants, the general case where $0 < h < 1$ may finally be evaluated. Once again, dividing rate equations 4.22b by 4.22a, and utilizing the unit activity for W, as well as the activities in equation 4.23, gives

$$R_E = (k_3 P_v^* - k_4 R_w)/(k_1 P_v - k_2) \tag{4.28a}$$

At this point, only algebraic manipulation is required to eliminate the k values, and to translate the last expression into a useful form. Substituting in the relationships from equations 4.22a and b,

$$R_E = (k_3/k_1)(P_v^* - P_v^{*0} R_w)/(P_v - P_v^0) \tag{4.28b}$$

Now, recognizing that the humidity is equal to the ratio,

$$h = P_v/P_v^0 \tag{4.29}$$

and using equations 4.23 and 4.27c, and rearranging, gives

$$R_E = \frac{R_w - \alpha_{eq} h R_v}{\alpha_{evap}^0 (1 - h)} \tag{4.30}$$

This result allows calculation of the isotope ratio R_E of the H_2O that, in effect, is being lost from the water body at any instant. Note that this ratio represents the net, weighted difference between the incoming and outgoing transport rates, rather than some physical sample that can be collected and analyzed. It is even more useful to express this last result in terms of a fractionation factor, here termed α_{evap}, between the lost component and the water body, at any instant. This may be achieved by simply dividing both sides of the equation by R_w, and then reciprocating:

$$\alpha_{evap} = \frac{R_w}{R_E} = \frac{\alpha_{evap}^0 (1 - h)}{1 - \alpha_{eq} h R_v/R_w} \tag{4.31}$$

Again, α_{eq}, α_{evap}^0, and α_{evap}, respectively, refer to the isotopic fractionation factors between water and water vapor at equilibrium ($h = 1$), and to nonequilibrium fractionation factors between the water and the net component lost at any instant under zero humidity ($h = 0$) conditions and under the actual humidity h of the process in question.

This important result allows many evaporation processes to be modeled. Several examples are given below. An equation of this form was originally developed from diffusion theory by Ehhalt and Knott (1965); the kinetic derivation provided above was developed completely independently of that work, with completely different starting equations. It is interesting to note that both theories give very similar results, and that very few assumptions are required to develop equation 4.31 from kinetic theory. This result arises naturally from very simple definitions, combined with rate equations 4.22a and b that simply assert that a direct proportionality must exist between the forward and reverse reaction rates and the concentrations of the relevant isotopic species in the water and water vapor.

4.3.1 Case 1: Evaporation into a Dry Atmosphere

Laboratory experiments illustrate the process of evaporation into completely dry tank air. This case may be evaluated with the Rayleigh fractionation equation, which, for this case, is written as

$$d \ln R_w = (1/\alpha_{evap}^0 - 1)d \ln f \tag{4.32}$$

At zero humidity, α_{evap}^0 is a constant, so equation 4.28 may be directly integrated to give

$$R_w = R_w^i f^{u^0} \tag{4.33a}$$

where

$$u^0 = 1/\alpha_{evap}^0 - 1 \tag{4.33b}$$

as was basically argued by Craig et al. (1963). Here, the quantity f represents the mass fraction of water that remains at any time, subsequent to the initiation of evaporation into a completely dry atmosphere.

The quantity α_{evap}^0 may be experimentally determined by simply plotting $\ln(1000 + \delta)$ versus $\ln f$ for such experiments. Typical values for this factor at 15°C are 1.106 and 1.027 for hydrogen and oxygen isotopes, respectively (see below).

4.3.2 Case 2: Evaporation into an Infinite, Humid Atmosphere

Several workers have studied the evaporation of water in shallow pans into the atmosphere. Under ordinary conditions, the value of α_{evap} will change continuously during the course of an evaporation process. For this case, the Rayleigh equation is

$$d \ln R_w = (1/\alpha_{evap} - 1) d \ln f \tag{4.34}$$

Equation 4.31 may be inserted into equation 4.34 and the result integrated with standard methods, to give the following (Stewart, 1975):

$$\boxed{\left(\frac{R_w - R_w^s}{R_w^i - R_w^s}\right) = f^u} \tag{4.35a}$$

where the exponent u is

$$\boxed{u = \frac{1 - \alpha_{evap}^0(1 - h)}{\alpha_{evap}^0(1 - h)}} \tag{4.35b}$$

Here, R_w, R_w^i, and R_w^s refer to the isotopic ratio of the water at any time, at the initial time, and at long times when the evaporation has gone nearly to completion. For any constant atmospheric condition, the value of R_w^s is a constant:

$$R_w^s = \frac{\alpha_{eq} h R_v}{1 - \alpha_{evap}^0 (1 - h)}$$

(4.35c)

Note that equations 4.35a, b, and c correctly reduce to equations 4.33a and b for the case of zero humidity.

The ratio R_w^s is a very important parameter. At saturation, where $h = 1$, R_w^s is clearly equal to $\alpha_{eq} R_v$, representing the isotopic ratio of the water that would be in equilibrium with the ambient atmospheric vapor. In contrast, under relatively high but unsaturated humidities, R_w^s is a "stationary state" ratio that the water approaches over time. In effect, as evaporation proceeds, the isotope ratio R_w of the water continuously adjusts until it approaches the value R_w^s, which in this case is equal to the ratio R_E of the net component of water that is lost at any instant. After this condition is achieved, the water continues to evaporate, and f accordingly decreases, but there is no further change in R_w because α_{evap} has become equal to unity. A typical evaporation curve illustrating this behavior is shown in Figure 4.6.

Still other behaviors are realized at lower humidities. The relationships are most easily demonstrated by a set of curves for equations 4.35a, b, and c, where all parameters are held constant except for the humidity, which is varied over a broad range (Figure 4.7). Note that the curves have a concave-down form, and that the "stationary state" behavior described earlier is realized for relatively high values of h, as shown by the curves for $h = 0.9$ to $h = 0.6$ (i.e., for humidities of 90–60%). However, for lower values of h, the curvature changes to concave up, and no true "stationary state" is realized, even though equation 4.35c still describes the final value of the last bit of water, when evaporation is complete.

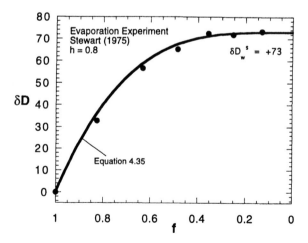

Figure 4.6 Graph of δD values versus f for a water drop that is undergoing evaporation at a temperature of 16.4°C. The relative humidity of 80% and the δD_a value of $-39.3‰$ of vapor in the external atmosphere are all that is needed to define the value of $+73$ for the isotopic stationary state, and to define the entire calculated curve, utilizing the equations 4.35a, b, and c, and 4.56b, and 4.66. Experimental data (circles) are from Stewart (1975).

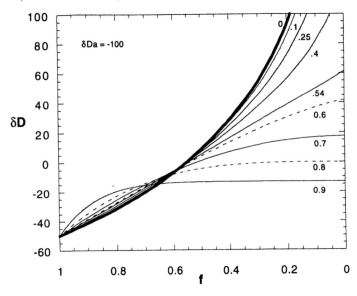

Figure 4.7. Graph of δD values versus f for an evaporating water body at 20°C for the different humidities indicated by the small numbers, all calculated with equation 4.35a, b, c. The δD value of the initial water is -50 in each case, while the ambient vapor is $-100‰$. For relatively high humidities the concave-down curves approach a flat, "stationary-state" at low f, where further evaporation reduces volume while producing negligible isotopic change. However, for a humidity of 54% the trend is linear, while for still lower humidities concave-upward trends are realized. For humidities less than 8.5%, true Rayleigh behavior is exhibited, with the δD values increasing without bound as f approaches zero.

An interesting case between these "concave-up" and "concave-down" behaviors is realized when the value of u equals unity, in which case δ_w varies linearly with f. According to equation 4.35b, this situation occurs when

$$h = 1 - 0.5/\alpha^0_{evap} \qquad (4.36a)$$

which is realized when h is approximately 0.54.

Another case of interest occurs for the situation where u equals zero, which equation 4.35b indicates occurs when

$$h = 1 - 1/\alpha^0_{evap} \qquad (4.36b)$$

The indicated value of h for deuterium is approximately 0.085. For humidities lower than this, the calculated values for R^s_w are not real, and the δD values instead increase without bound as f becomes very small. As a result, the evaporation trends approximate the Rayleigh behavior that is realized at zero humidity (Figure 4.7).

According to this reasoning, the heavy isotope content of water will normally increase during the course of evaporation. However, this result is not inevitable, because the process depends not only on the humidity, but also on the isotope ratios of the ambient atmospheric vapor and the initial water. Thus, the isotope ratio of a water body can actually decrease during evaporation if the initial water is highly enriched in heavy isotopes relative to the vapor. It is also possible for ^{18}O enrichment to accompany D depletion, or vice versa. It is therefore advantageous to monitor evaporation conditions by using two or more pans of water, particularly if one is considerably enriched in heavy isotopes. In this case, both waters attempt to attain the same steady-state value, but approach it from opposite directions. Thus, some water bodies progressively increase in heavy isotopes, while the others becomes progressively depleted, until the common value of the steady state is realized (Figure 4.8)

4.3.3 Slope of δD versus $\delta^{18}O$ Trends for Evaporated Waters

For natural water bodies, it is more useful to study the mutual variations of the δD and $\delta^{18}O$ values of the water samples, rather than the trends of either value as a function of the parameter f, simply because the isotopic data are easily gathered for any given sample. In this regard, it is commonly asserted that evaporated waters in nature are typically enriched in heavy isotopes, and that they typically display a slope of about 5 on a standard plot of δD versus $\delta^{18}O$ values. The following paragraphs show that no characteristic slope can be defined for evaporated samples.

The process of evaporation does indeed produce nearly linear trends on a δD versus $\delta^{18}O$ graph. Figure 4.9 shows a simple experiment where a beaker of water

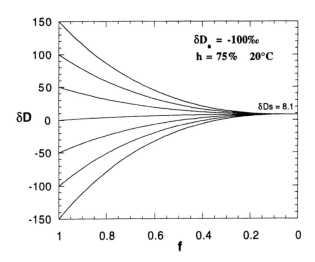

Figure 4.8 Graph of δD values versus f for isotopically distinct, 20°C waters that evaporate into an atmosphere with a humidity of 75% and a δD value of $-100‰$. The initial δD values of the waters range from -150 to $+150$, and are indicated by the values at $f = 1$. In all cases the waters exponentially approach an isotopic "stationary state" of $+8.1$. Waters with initial values lower than $+8.1$ become enriched in deuterium during evaporation, while waters that have high initial values become depleted. All curves were calculated with equation 4.35a, b, c.

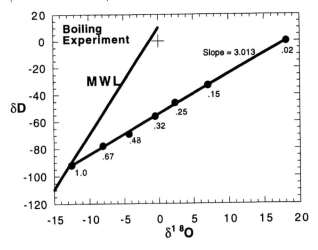

Figure 4.9 Graph of δD versus $\delta^{18}O$ values for the evaporation of boiling water into a 22°C atmosphere with δD and $\delta^{18}O$ values of -18.0 and -120‰, respectively. Numbers are f values. The humidity of 37% corresponds to a humidity of only 1% when referred to the water temperature. The slope of the evaporation trend is only 3.013, and is very similar to that exhibited for acid-sulfate waters (e.g., Figure 3.4). Data from Ingraham et al. (1991).

was placed on a hotplate and successively sampled as it boiled away. Under these conditions, the observed relationship is essentially a straight line with a slope of approximately 3.

Determining the theoretical δD versus $\delta^{18}O$ slope of an evaporation trend from the above relationships is more difficult than might appear. In fact, there is no simple relationship that relates such slopes to the humidity. This problem can be elucidated by rewriting equation 4.34 in terms of the above-defined parameters to reformulate the Rayleigh equation for evaporation (R. E. C., unpublished):

$$dR_w = u(R_w - R_w^s)\,d\ln f \qquad (4.37a)$$

or

$$dR_w = -(R_w - R_E)\,d\ln f \qquad (4.37b)$$

Equation 4.37a is very useful; note that equation 4.35a follows from it directly. If two equations of this form are written, one for the hydrogen and one for the oxygen isotopes, and these relationships are divided and the result converted to δ-values, the slope of an evaporation trend follows directly:

$$\frac{d\,\delta D_w}{d\,\delta^{18}O_w} = \frac{u_D(\delta D_w - \delta D_w^s)}{u_{ox}(\delta^{18}O_w - \delta^{18}O_w^s)} \qquad (4.38a)$$

This result is almost perfectly linear, and to a very good approximation the slope is given by the equation

$$\frac{d\,\delta D_w}{d\,\delta^{18}O_w} \cong \frac{(\delta D_w^i - \delta D_w^s)}{(\delta^{18}O_w^i - \delta^{18}O_w^s)} \qquad (4.38b)$$

where δD_w^i and $\delta^{18}O_w^i$ represent the initial values of the water, prior to any evaporation. Now, it is apparent why there is no simple value for the slope. For any given atmospheric condition, represented by the humidity and the isotopic ratios of the vapor, there will be predetermined values of δD_w^s and $\delta^{18}O_w^s$ that the evaporating water bodies will attempt to attain. Those values will be entirely independent of the initial composition of the water bodies. On a graph of δD versus $\delta^{18}O$ values, the slope of the evaporation trend may take on any value for a single set of conditions, depending on the relative positions of the two points that represent the initial water and the "stationary-state" value. It is analogous to the situation where the direction to St. Louis depends on where you are traveling from.

A useful graphical technique for the slope is as follows. For a given atmosphere that has vapor with specified δD and $\delta^{18}O$ values, the δD and $\delta^{18}O$ values of the water that would be in equilibrium with that vapor are easily defined by the equilibrium fractionation factor. These values are the same as the values for δD_w^s and $\delta^{18}O_w^s$ at $h = 1$. For any lower value of the humidity, equation 4.35c may be used to calculate the values for δD_w^s and $\delta^{18}O_w^s$. These values may be precisely defined by that equation, but typically plot along a slope of about 3.5 that extends from the equilibrium ($h = 1$) point. The trajectory followed by any given water that is undergoing evaporation will be very close to a straight line that connects the initial value with the "stationary-state" point appropriate for the particular isotope ratios and humidity of the atmosphere. This trajectory can have any slope; an example is shown in Figure 4.10.

One useful generalization can be made for the condition that the isotope ratios of the vapor happen to be close to the equilibrium ratios, independently of the humidity. In other words, suppose that for both hydrogen and oxygen isotopes,

$$\alpha_{eq} R_v \cong R_w \qquad (4.39a)$$

For this case, it may be shown that the slope of an evaporation trajectory is approximately

$$\frac{d\,\delta D_w}{d\,\delta^{18}O_w} \cong \frac{(1 - 1/\alpha_{evap,D}^0)(1000 + \delta D_w)}{(1 - 1/\alpha_{evap,ox}^0)(1000 + \delta^{18}O_w)} \qquad (4.39b)$$

A typical value would be 3.5.

4.3.4 Case 3: Evaporation of Salt Water into an Infinite, Humid Atmosphere

Stewart (1975) studied the evaporation of suspended seawater drops into the atmosphere. The results have an interesting form that has not been explained (Figure 4.11). However, equation 4.34 may be used to examine this situation, simply by recognizing that the salt content of the drop will necessarily increase as evaporation proceeds. This will necessarily lead to progressive changes in the activity of the water in the drop. It can be assumed that the situation is well

Figure 4.10 Graph of calculated δD versus $\delta^{18}O$ showing the trend (heavy line, slope 3.76) of isotopic stationary states associated with a particular atmospheric vapor (open circle). Values of humidity are indicated by the small numbered "ticks" along the trend line; the value at $h = 1$ (i.e., 100%) represents the water that would be in chemical and isotopic equilibrium with the ambient vapor. The isotopic shift that an arbitrary water body would undergo while evaporating into this atmosphere may be estimated by simply connecting its initial composition with the appropriate stationary state. Such trends are nearly linear, as shown for the evaporation trend (dashed line) for the water (square) that is evaporating into an atmosphere of 60% humidity.

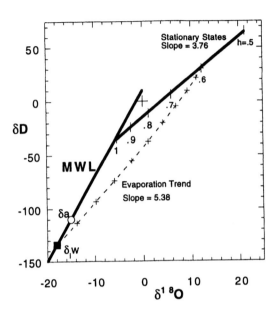

approximated by a simple NaCl solution that has an initial salinity of 3.5 wt. %. Near 20°C, the data for the saturated vapor pressure P_s of aqueous NaCl solutions (*International Critical Tables*, 1933) may be approximated as a function of the weight percent (s) of salt as

$$P_s = P^0(1 - 0.00407s - 0.000187s^2) \qquad (4.40a)$$

where P^0 is the vapor pressure of pure water at the temperature of interest. Moreover, the effective humidity must be referred to the saturated vapor pressure above the salt solution:

$$h_{eff} = hP^0/P_s \qquad (4.40b)$$

The equilibrium fractionation factors also vary as a function of the salt content. An empirical equation for this effect on the hydrogen isotope fractionation factor is (Stewart and Friedman, 1975)

$$\alpha_{eq}^* = \alpha_{eq}(1 - 0.002M) \qquad (4.40c)$$

where α_{eq}^* is the equilibrium fractionation factor between the NaCl solution and water vapor, α_{eq} is the equilibrium fractionation factor for pure water, and M is the molality of NaCl.

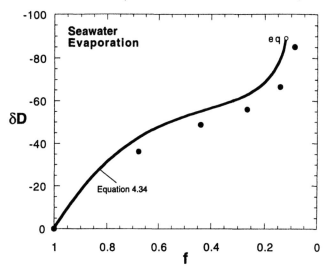

Figure 4.11 Graph of calculated δD values versus f for a seawater drop that is undergoing evaporation at a temperature of 16.5°C. The calculated curve represents the numerical integration of equation 4.34, utilizing the relative humidity of 80% and the δD_a value of $-147.8‰$ of vapor in the external atmosphere, and incorporating corrections for changes in activity and fractionation factor with increasing salinity (R.E.C., unpublished, 1976). Under these circumstances, the drop does not evaporate to completion, but rather increases in salinity until it achieves a point (small open circle) representing chemical and isotopic equilibrium with the external atmosphere. Experimental data (filled circles) are from Stewart (1975).

All of these parameters change as the evaporation proceeds, because the salt content of the drop increases progressively. As a result, differential equation 4.34 may still be used to evaluate this process, but the effective humidity, the kinetic fractionation factor α_{evap}, and the equilibrium fractionation factor α_{eq}^D must all be expressed as a function of f. Once this is done, a numerical integration of all the equations may be conducted, with all the parameters being evaluated and changed during each incremental step of the process. The calculated curve shown in Figure 4.11 agrees very well with Stewart's data, and, in particular, reproduces the peculiar inflection point observed at $f \sim 0.3$.

A very interesting feature is that the calculation does not progress to the point $f = 0$, where the drop has disappeared, but rather approaches a limiting value at $f \sim 0.12$. The mathematical limit of the equations was examined by Prof. F. Bohnenblust (pers. comm., 1976). His result indicates that, at this final point, the activity of water in the highly saline drop had been reduced until it was equal to the actual humidity, so that the effective humidity became 100%. At this point, the evaporation ceased, and the drop in this case attained both chemical and isotopic equilibrium of the drop with the vapor. From the final experimental point, an equilibrium fractionation factor of 1.074 may be calculated that is in reasonable agreement with the data of Stewart and Friedman (1975; equation 4.40c).

4.3.5 Case 4: Marine Vapors

An important application of the evaporation equations is to model the isotope ratios of atmospheric water near the sea–air interface. This problem differs from the examples given earlier, because the system is clearly not akin to a small water body being dominated by a large atmosphere. As a result, in this case the isotope ratios of the water will not vary appreciably over time, but instead will remain constant. In such a situation, one might expect the atmosphere to simply attain chemical ($h = 1$) and isotopic equilibrium with the water, but atmospheric turbulence and transport processes generally prevent this condition from being realized.

The actual situation is best visualized as one of "steady state." Under these steady conditions, even though evaporation proceeds uninterrupted, the humidity and the isotopic ratios of the water and the atmospheric vapor all remain constant over time. Any water lost from the ocean is simply transported away in the atmosphere, recharging the hydrologic cycle, and is ultimately destined to return to the ocean either as direct precipitation or meteoric runoff from the continents.

While this scenario might seem complicated, the mathematical constraints of this steady-state system are actually quite simple. The entire process may be evaluated with equation 4.31, as follows. Craig and Gordon (1965) estimated that the average meteoric precipitation for Earth has a $\delta^{18}O$ value of -4.5 and a δD value of -26; these values must reflect the isotopic ratios (R_E) of water lost from the ocean. Given that the $\delta^{18}O$ and δD values of ocean water are both very close to zero, the indicated values of α_{evap} are therefore close to 1.0045 and 1.0267, respectively. Equation 4.31 may then be written in a form that allows the δ-values of the vapor to be calculated for any steady-state humidity. That is, the δ-values of vapors associated with SMOW may be found from

$$\alpha_{evap} = \frac{\alpha^0(1 - h)}{1 - \alpha_{eq}h(1 + \delta_v/1000)} \tag{4.41a}$$

Table 4.1 provides example calculations for this simple model, and Figure 4.12 shows a good correspondence between these values and the $\delta^{18}O$ and δD values of marine vapors measured by Craig and Gordon (1965). The trend of computed values is approximately

$$\delta D_v = 2.85\,\delta^{18}O_v - 50.7 \tag{4.41b}$$

This trend intersects the MWL at a $\delta^{18}O$ value of about -11.8, which occurs at a humidity of about 91%.

4.4 Lake Balance

Evaporation is a key transport process in the hydrology of lakes, particularly in arid regions. A standard hydrologic balance of the volume of a lake expresses the difference between the volumetric input rates (first parenthesis on the right-

Table 4.1 Calculated $\delta^{18}O$ and δD Values of Marine Vapor as a Function of Humidity (Equation 4.41a)

h	$\delta^{18}O$	δD
1.00	−9.7	−78.3
0.95	−10.8	−81.5
0.90	−12.1	−85.1
0.85	−13.4	−89.0
0.80	−15.0	−93.4
0.75	−16.8	−98.5
0.70	−18.8	−104.2

Assumed values: $\alpha_{evap} = (1.0045, 1.0267)$; $\alpha_{eq} = (1.00979, 1.0850)_{20°C}$; $\alpha_{evap}^0 = (1.0260, 1.094)_{20°C}$; $\delta D_w = 0$; $\delta^{18}O_w = 0$.

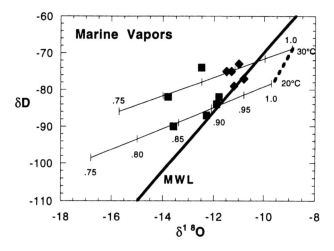

Figure 4.12 Graph of observed and calculated δD versus $\delta^{18}O$ values of marine vapors. The observed vapors, from Craig and Gordon (1965), are from the North Pacific Ocean at 0–14°N (diamonds) and 14–32°N latitudes (squares). Vapors in equilibrium with SMOW would lie along the dotted line. Steady-state vapors at humidities of 100–75% would have lower δ-values, as indicated by the calculated trends (equation 4.41) for evaporation at 20°C and 30°C, assuming that the net vapor that is removed has the values of −4.5 and −26.0 suggested by Craig and Gordon (1965). The isotopic difference between observed and equilibrium marine vapors explains the "deuterium excess" of the MWL.

hand side) and the output rates (second parenthesis on the right-hand side) of the water:

$$dV/dt = (I + P) - (O + E) \qquad (4.42a)$$

where dV/dt is the change in the volume of water with time, I is the volumetric rate of stream and groundwater inflow, P is meteoric precipitation that falls directly on the surface of the lake, O is stream and groundwater outflow, and E is evaporation from the lake surface (e.g., Gonfiantini, 1986).

Equation 4.42a may be viewed as the conservation relation for the abundant isotopic species of water, $H_2{}^{16}O$. A corresponding equation can be used to express mass conservation for the trace isotope, either HDO or $H_2{}^{18}O$, in the different water components:

$$d(VR_w)/dt = IR_I + PR_P - OR_O - ER_E \qquad (4.42b)$$

where the R values represent the isotope ratios of the component indicated by the subscript, and R_w is the average ratio of the lake water. If the lake is well mixed, then the isotopic ratio of any outflowing water, R_O, will be necessarily equal to the average lake water, R_w, as assumed below. Figure 4.13 illustrates the isotopic fluxes associated with a generalized lake.

In transient problems, either or both of the volume and isotope ratio of the lake may change over time. It is also possible for the lake to attain a steady state, in which case dV/dt and $d(VR_w)/dt$ are both zero, and the level and the isotope ratio of the lake do not vary over time. In this important case of steady state, equations 4.42a and b become

$$I + P = O + E \qquad (4.43a)$$

and

$$IR_I + PR_P = OR_w + ER_E \qquad (4.43b)$$

Three simple cases illustrating the use of the latter equations are given in sections 4.4.1–4.4.3.

4.4.1 Case 1: Throughflow Lake

One common type of lake in temperate regions is a "throughflow" lake, in which inflowing water simply flows out. If the lake is small and the flow rate relatively large, direct precipitation and evaporation may be neglected. In the case of steady state, the lake volume is constant, and the isotope ratios of the inflow, outflow, and lake are all constant and identical. Thus,

$$I = O \qquad \text{for } E, P \sim 0 \qquad (4.44a)$$

Figure 4.13 Diagram of isotopic fluxes associated with a lake. The isotope ratios for the water (R_w), atmosphere water vapor (R_a), vapor (R_w/α) in the boundary layer, stream inflow plus direct precipitation (R_{in}), and lake outflow (R_{out}) are illustrated. The wavy arrows schematically illustrate net evaporation that would be accompanied by net loss of a component that has ratio R_E. See text.

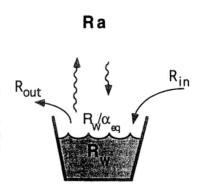

and

$$R_I = R_O = R_w \tag{4.44b}$$

Different cases of transient problems may also be treated. For example, for a well-stirred lake with a spillway that maintains a constant volume V, the inflow and outflow rates are equal. If the initial isotope ratio of the lake (R_w^i) is affected by a large variation in the isotope ratio of the inflow, as might occur following a large storm with a very unusual isotope ratio, then the isotope ratio of the lake would recover as follows:

$$\frac{R_w - R_I}{R_w^i - R_I} = e^{-It/V} \tag{4.44c}$$

If the isotope ratio and volume of the lake are both changing, equations 4.42a and b must be solved simultaneously.

4.4.2 Case 2: Simple Closed Basin

For the most simple type of hydrologically closed basin, all inflowing water is removed by evaporation. Many terminal lakes in arid regions essentially achieve this condition, because they adjust their area over nearly level playas until the evaporation rate becomes equal to the inflow rate. For this situation, a steady state may be achieved, where the inflow rate equals the evaporation rate, so $I = E$. At steady state, it is necessary then for R_I to be equal to R_E. In other words,

$$\alpha_{evap} = \frac{R_w}{R_I} \tag{4.45a}$$

If the above condition is introduced into equation 4.31, the steady-state isotope ratio of the closed lake is seen to be

$$\boxed{R_w = \alpha_{evap}^0 (1 - h)R_I + \alpha_{eq}hR_v} \tag{4.45b}$$

where the quantities were all defined previously. Note that this result is different than R_w^s. Consequently, the isotope ratio of a terminal lake with inflow can differ considerably from the "stationary-state" value attained in evaporation pans on its shore.

To convert this result to δ-values, each R need only be replaced by the quantity $1000 + \delta$, with care being taken to retain the appropriate subscripts (equation A3.2.1). Moreover, for a basin with significant direct precipitation, the effect of the rain may be incorporated by simply adjusting the effective "inflow" ratio, R_I, to reflect the volumetric proportions of the stream inflow and precipitation components.

4.4.3 Case 3: Throughflow Lake with Evaporation

In a more complex lake, a steady condition may also be achieved if the inflow is matched by a combination of evaporation and outflow. For this situation, where $I = E + O$, it is useful to define the ratio X_E:

$$X_E = \frac{E}{I} \qquad (4.46a)$$

This ratio represents the volumetric fraction of inflowing water that is lost to evaporation. This ratio will be related to the concentration of salts in the lake, C_w, relative to that of the inflow C_I, in the simplest conditions by

$$C_I = C_w(1 - X_E) \qquad (4.46b)$$

Of course, if minerals such as halite are precipitated, this expression will not be correct, although the indicated value for X_E will be close enough to unity for such cases that this limitation is unimportant.

For a well-mixed lake, where the isotope ratio of the outflow is the same as the average lake water,

$$R_w = \frac{\alpha_{evap}^0 (1 - h)R_I + \alpha_{eq}hX_E R_v}{X_E + \alpha_{evap}^0 (1 - h)(1 - X_E)} \qquad (4.46c)$$

To convert this result to δ-values, each R need only be replaced by $1000 + \delta$, retaining the appropriate subscripts. Note that this result correctly reduces to equation 4.45b for the case where X_E approaches unity, and to the condition $R_w = R_I$ when X_E equals zero. An interesting application would be to a chain of desert lakes.

4.5 Isotopic Flux and Soil Evaporation

4.5.1 Basic Equations

Several processes of chemical transport are a function of position as well as time, so that spatial gradients in concentrations or isotope ratios are important. For such cases, the laws of diffusion are more usefully employed than are those of elementary kinetics. The examples examined here will include evaporation from standing water bodies and from soils.

4.5.1.1 Fick's Laws   Fick's laws provide the fundamental basis for the theory of diffusion. Fick's first law establishes that the flux of material F is directly proportional to the spatial gradient in concentration C:

$$F = -D\nabla C \qquad (4.47a)$$

where D is the diffusion coefficient, which has the cgs units cm^2/s. The flux therefore carries the cgs units of $g/cm^2\text{-}s$, and represents the amount of material that passes across a 1-cm^2 reference plane in a second. The negative sign indicates that the transport occurs in a direction opposite that of increasing concentration.

Fick's second law may be derived by taking the divergence of equation 4.47a, and recognizing that the divergence of the flux is necessarily related to a change in the concentration in the reference volume with time. For cases where the diffusion coefficient is constant, the result is

$$\partial C/\partial t = D\nabla^2 C \qquad (4.47b)$$

where ∇^2 is the Laplacian operator. The preceding equations are all very general because the gradient, the Laplacian, and the divergence may be expressed in cartesian, cylindrical, or spherical coordinates. The simple case of one-dimensional transport is most important, and, though less general, will be emphasized here (boxed equations). For this situation, the flux becomes

$$\boxed{F = -D\,\frac{\partial C}{\partial x}} \qquad (4.48a)$$

and the second law becomes

$$\boxed{\frac{\partial C}{\partial t} = D\,\frac{\partial^2 C}{\partial x^2}} \qquad (4.48b)$$

Several excellent books present solutions for diffusion problems (e.g., Crank, 1975). Furthermore, because the mathematical form of the equations is basically the same, books on heat conduction also provide additional useful solutions (e.g., Carslaw and Jaeger, 1976). More complex cases—for example, where the diffusion coefficient is a variable or is anisotropic, or for volumes that contain a source or a sink of concentration—are also treated in these books. Such complications are not relevant to the examples given in the following.

4.5.1.2 Diffusion–Advection Equation   If advective transport of material accompanies diffusion, additional terms need to be added to the two diffusion equations. The "first law" for the flux is modified to

$$F = -D\nabla C + VC \qquad (4.49a)$$

where V is the velocity of the flow, and the product VC then represents the flux of material due to advection. The modified "second law" may be determined by taking the divergence of the last expression. For the case of incompressible flow, the divergence of the velocity is zero, and, if D is also constant, the "second law" becomes

$$\partial C/\partial t = D\nabla^2 C + V\nabla \cdot C \qquad (4.49b)$$

where $\nabla \cdot C$ is the divergence of the concentration.

For one-dimensional transport, this pair of equations may be simplified. The flux equation becomes

$$F = -D\frac{\partial C}{\partial x} + VC \tag{4.50a}$$

while the equation for diffusion and incompressible flow becomes

$$\frac{\partial C}{\partial t} = D\frac{\partial^2 C}{\partial x^2} + V\frac{\partial C}{\partial x} \tag{4.50b}$$

4.5.1.3 Steady State   For the important case of steady state, spatial gradients may exist but there are no changes in concentration with time. For this situation, the various forms of the second law may be considerably simplified. For example, equation 4.47b reduces to Laplace's equation,

$$\nabla^2 C = 0 \tag{4.51a}$$

or, for one-dimensional diffusion,

$$\frac{\partial^2 C}{\partial x^2} = 0 \tag{4.51b}$$

For steady-state conditions in a diffusion–advection situation, equation 4.49b becomes

$$D\nabla^2 C = -V\nabla \cdot C \tag{4.52a}$$

or, for one-dimensional advective–diffusive transport,

$$D\frac{\partial^2 C}{\partial x^2} = -V\frac{\partial C}{\partial x} \tag{4.52b}$$

4.5.1.4 Problem-Solving Methodology   An inspiring effort has been expended over the centuries to solve the preceding and related equations. Numerous excellent books treat these aspects in detail, making further development pointless here. Attention can be confined to translating the available mathematical results into the language of isotope distribution.

The mathematical solutions to differential equations must satisfy two different kinds of constraints. First, the solution must satisfy the original differential equation. This may generally be tested by differentiating the solution, inserting the result into the original differential equation, and determining whether a mathematical identity can be obtained. Second, the solution must satisfy any initial conditions or boundary conditions that are proscribed for the given problem, which may be determined by inspection. In the following examples, the relevant differential equation is identified, the boundary and/or initial conditions are stated, and the solution is simply provided but not developed.

One very useful method of proceeding is to first identify appropriate published equations that give concentration as a function of distance and time, for the particular conditions of interest. The flux may be determined from those concentration equations by differentiation; for example, by using equation 4.48a. Next, two analogous flux equations are written, one for the common isotope and one for the trace isotope, and these equations are then divided and translated in terms of isotope ratios and other parameters of interest. Several examples of this and other techniques will be given here.

4.5.2 Evaporation

The processes of evaporation may be usefully quantified with diffusion theory. The most straightforward case, originally evaluated by Ehhalt and Knott (1965), involves water vapor transport into the atmosphere, and away from a stationary planar surface of water that has a constant composition. The initial and boundary conditions for the atmospheric half-space are

Initial condition: at $t = 0$, $C = C_i$ for all $x > 0$ (4.53a)

Boundary condition: at $x = 0$, $C = C_{sfc}$, a constant   (4.53b)

The well-known solution to differential equation 4.48b for this geometry and the stated conditions is

$$\frac{C - C_{sfc}}{C_i - C_{sfc}} = \text{erf}\left[\frac{x}{\sqrt{4Dt}}\right] \tag{4.53c}$$

where erf is the error function, and its argument, $x/\sqrt{4Dt}$, is dimensionless. The flux may be found by differentiating this result, per equation 4.48a, to obtain

$$F = (C_i - C_{sfc})\sqrt{\frac{D}{\pi t}} \exp\left[\frac{-x^2}{4Dt}\right] \tag{4.53d}$$

Evaluating this result at the water–atmosphere interface ($x = 0$) gives

$$F_{x=0} = (C_i - C_{sfc})\sqrt{\frac{D}{\pi t}} \tag{4.54a}$$

The problem at hand reduces to translating this equation into the isotope ratios and other parameters of interest. This may be accomplished by recognizing that a companion for each of the preceding equations may be written to represent the concentration of the trace isotope of interest. In particular, the companion to equation 4.54a is

$$F^*_{x=0} = (C^*_i - C^*_{sfc})\sqrt{\frac{D^*}{\pi t}} \tag{4.54b}$$

where the asterisks on the flux and the concentrations denote the trace isotope, and D^* represents the rate of diffusion of the trace isotope in the atmosphere. Now, dividing 4.54b by 4.54a gives a result that, most interestingly, is independent of the time:

$$\left(\frac{F^*}{F}\right)_{x=0} = \frac{(C^*_i - C^*_{sfc})}{(C_i - C_{sfc})}\sqrt{\frac{D^*}{D}} \tag{4.54c}$$

The ratio F^*/F is clearly equal to the familiar quantity R_E, representing the isotopic ratio of the net component of H_2O removed from the water body at any time. In addition, the following conditions and variable assignments are appropriate:

$$R_E = \left(\frac{F^*}{F}\right)_{x=0} \tag{4.55a}$$

$$R_v = \left(\frac{C_i^*}{C_i}\right) \tag{4.55b}$$

$$R_{v,sfc} = \left(\frac{C_{sfc}^*}{C_{sfc}}\right) = \frac{R_{w,sfc}}{\alpha_{eq}} \tag{4.55c}$$

and

$$h = \left(\frac{C_i}{C_{sfc}}\right) \tag{4.55d}$$

Equation 4.55a states that the isotope ratio R_E of the material evaporated from the water body must necessarily equal the ratio of the fluxes for the two isotopes. Equation 4.55b indicates that the isotopic ratio of the atmospheric vapor at distance, here considered to be an enormous reservoir that is basically unaffected by the evaporation, is constant and equal to the ratio of initial concentrations of H_2O and H_2O^* in the vapor. However, in the boundary layer immediately above the air–water interface, it can be assumed that both chemical and isotopic equilibrium of the vapor and the underlying water are achieved. As a result, the isotopic ratio of the vapor in this layer is given by equation 4.55c. In addition, equation 4.55d defines the humidity of the atmosphere at distance, and basically asserts that a saturated vapor that has composition C_{sfc} exists in this boundary layer.

If these quantities are introduced into the equation for the flux ratio equation (4.54c) and the result simplified, the equation of Ehhalt and Knott (1965) is obtained:

$$\alpha_{evap} = \frac{R_{w,sfc}}{R_E} = \frac{\alpha_{eq}\sqrt{\dfrac{D}{D^*}}(1-h)}{1 - \alpha_{eq}hR_v/R_{w,sfc}} \tag{4.56a}$$

This result has a stunning similarity to equation 4.31, which was independently developed from simple kinetic theory. The principal difference is that diffusion theory also provides an explanation for the quantity α_{evap}^0, representing the separation factor at zero humidity. That is, at zero humidity, equation 4.56a becomes

$$\alpha_{evap}^0 = \alpha_{eq}\sqrt{\frac{D}{D^*}} \tag{4.56b}$$

Stewart (1975) provides additional discussion and experimental data regarding this important factor.

4.5.3 Evaporation: Composite Medium

The results of Ehhalt and Knott (1965) may be extended into a more complicated geometry)—that of the infinite composite medium—comprising two half-spaces of air (region 1) over liquid water (region 2), separated by a planar, stationary interface. For this situation, the initial and boundary conditions are as follows:

$$\text{Initial condition:} \quad \text{at } t = 0, C_1 = C_{1,i} \text{ for all } x > 0 \quad (4.57a)$$

$$\text{and} \quad \text{at } t = 0, C_2 = C_{2,i} \text{ for all } x < 0 \quad (4.57b)$$

$$\text{Boundary condition:} \quad \text{at } x = 0, D_1\, \partial C_1/\partial x = D_2\, \partial C_2/\partial x \quad (4.57c)$$

$$\text{and} \quad \text{at } x = 0, C_2/C_1 = K \quad (4.57d)$$

The initial conditions establish the concentrations at the beginning of the problem, and for all time at large distance from the interface. The boundary conditions establish that the flux of material leaving one region must be equal to the flux of material entering the other, and additionally, that the water and water vapor are in an equilibrium relationship at the boundary layer at the interface, as controlled by the equilibrium constant K. The simultaneous solution to a pair of differential equations of the form of equation 4.48b, with one that represents region 1 and the other that represents region 2, for this situation and the stated conditions is well known. For region 1 (the air),

$$\frac{C_1 - C_{1,\text{sfc}}}{C_{1,i} - C_{1,\text{sfc}}} = \text{erf}\left[\frac{x}{2\sqrt{D_1 t}}\right] \quad (4.58a)$$

where the concentration at the interface ($x = 0$) is a constant:

$$C_{1,\text{sfc}} = \frac{C_{1,i} + C_{2,i}\sqrt{D_2/D_1}}{1 + K\sqrt{D_2/D_1}} \quad (4.58b)$$

For region 2, here representing the water, the solution is

$$\frac{C_2 - C_{2,\text{sfc}}}{C_{2,i} - C_{2,\text{sfc}}} = \text{erf}\left[\frac{|x|}{2\sqrt{D_2 t}}\right] \quad (4.58c)$$

where the concentration $C_{2,\text{sfc}}$ at the interface (i.e., at $x = 0$) is another constant, most compactly represented by the product of the equilibrium constant K and $C_{1,\text{sfc}}$; that is,

$$C_{2,\text{sfc}} = KC_{1,\text{sfc}} \quad (4.58d)$$

This result has many interesting ramifications. First, it is seen that the concentrations in the boundary layer are constants. As a result, equation 4.58a has the same form as 4.53c, so, by analogy, the result given by equation 4.56a must also provide the correct value for α_{evap} in this situation. Second, equation 4.58c predicts that vertical isotopic gradients could exist in the water body. For a trace isotope, these gradients would be distributed according to the error function. Rewriting equation 4.58c in terms of isotope ratios gives

$$\boxed{\frac{R_w - R_{w,sfc}}{R_{w,i} - R_{w,sfc}} = erf\left[\frac{z}{2\sqrt{D_w t}}\right]}$$

(4.58e)

where z is the water depth and D_w is the self-diffusion coefficient of the isotopic species of liquid water. This result demonstrates another major difference between diffusion theory and kinetic theory, in that the former commonly can directly provide information about spatial gradients in isotope ratios. Of course, if the water is stirred, then these small gradients will be eliminated.

One final complication is that for water bodies of finite depth or for geometries such as drops, the fluxes will necessarily vary with time. Equation 4.54c may therefore be viewed as a limiting condition, and, in particular, will be the exact solution for finite water volumes for short times.

4.5.4 Soil Evaporation: Saturated Zone

Evaporation from soils provides a fine example of the use of diffusion equations to study isotopic gradients. In effect, the boundary layer between the liquid water and the atmosphere is greatly magnified by the soil matrix. Water can therefore be collected and analyzed from soil samples taken at different depths, providing a detailed vertical profile of the isotopic gradients (e.g., Zimmermann et al., 1967; Barnes and Allison, 1988).

A realistic and important problem is the development of an isotopic profile in a saturated soil zone that extends from the surface (depth $z = 0$) to a constant source of replenishment at depth L. This replenishment may be viewed as a stationary "layer" of underlying water that provides water of constant composition $R_{w,res}$, at depth L, just as fast as water is removed at the surface by evaporation. The problem may then be examined in the case of steady state. For this circumstance, equation 4.52b is appropriate, and for a trace isotope species of liquid water the equation may be rewritten as

$$D\,\partial^2 R_w/\partial z^2 = -E\,\partial R_w/\partial z$$

(4.59a)

where the flow velocity has been replaced by the evaporation rate E, and where D now refers to the diffusion rate in the soil. The boundary conditions are

Boundary condition: at $z = 0$, $R_w = R_{w,sfc}$ (4.59b)

and   at $z = L$, $R_w = R_{w,res}$ (4.59c)

These conditions establish that, at depth L, the isotopic ratio of the soil water must be the same as that of the water below, and also, at steady state, the ratio at the soil–air interface must be a constant. The solution to equations 4.59a, b, and c is

$$\frac{R_w - R_{w,sfc}}{R_{w,res} - R_{w,sfc}} = \frac{1 - e^{-Ez/D}}{1 - e^{-EL/D}} \qquad (4.60a)$$

For the stated conditions, the isotopic variations have an exponential profile. If the zone of replenishment is very deep, this equation reduces to the relationship provided by Barnes and Allison (1988):

$$\frac{R_w - R_{w,res}}{R_{w,sfc} - R_{w,res}} = e^{-Ez/D} \qquad (4.60b)$$

The composition of the soil water at the surface—that is, $R_{w,sfc}$ in equations 4.60a and b—may be determined by the same arguments as were used to develop equation 4.45b. For this case, the result is

$$R_{w,sfc} = \alpha_{evap}^0 (1 - h) R_{w,res} + \alpha_{eq} h R_v \qquad (4.60c)$$

as has been argued by Barnes and Allison (1988). Experimental confirmation of such exponential isotopic profiles is given by Zimmermann et al. (1967) and Barnes and Allison (1988).

4.5.5 Soil Evaporation: Unsaturated Zone

Evaporation from unsaturated soils, above the water table, is a complicated process. Here, the transport of isotopes occurs by diffusion of both liquid water and water vapor phases, which coexist in the vadose zone. Moreover, different depth zones in the soil have different physical characteristics. A completely satisfactory theory for this complex situation has not yet been worked out. Here, and simplified after the arguments of Barnes and Allison (1983, 1988), the following assumptions are made: (1) the zone of interest is isothermal; (2) after substantial passage of time, the situation can be approximated by a "steady-state" isotope distribution; (3) in the uppermost layer of soil, above a depth z_{ef} representing the "evaporation front," the pore vapor is unsaturated and all transport of water molecules occurs in this vapor phase; and (4) the vapor phase is saturated ($h = 1$) at and below this evaporation front, and transport of water molecules is dominated by liquid-phase transport.

The uppermost zone is easiest to treat. In this zone, the isotopic transport is assumed to occur in the vapor phase. The humidity must vary with depth, notably from h_a, the humidity of the overlying atmosphere, at depth $z = 0$, to the saturation value of $h = 1$ at depth z_{ef} of the evaporation front. The relevant boundary conditions are then

Boundary condition:	at $z = 0$, $h = h_a$	(4.61a)
and	at $z = z_{ef}$, $h = 1$	(4.61b)

For steady-state conditions, the solution to Laplace's equation (4.51b) is $C = A + Bx$, where A and B are constants. Replacing the variable z for x, and calculating A and B from the boundary conditions of equations 4.61a and b gives

$$h_z = h_a + (1 - h_a)z/z_{ef}$$

(4.61c)

Thus, the humidity varies linearly with depth, as is required for steady state. Similar arguments may be made for the trace isotopic species of water vapor, which must also vary linearly. Thus,

$$P_z^* = P_a^* + (P_{ef}^* - P_a^*)z/z_{ef}$$

(4.61d)

where the asterisk denotes the heavy isotope of interest, and subscripts "a" and "ef" refer to the concentrations in the overlying atmosphere and at the evaporation front, respectively. The equation can be translated into isotope ratios with the following simple relationships:

$$h_a = P_a/P^0 \qquad h_z = P_z/P^0 \qquad R_a = P_a^*/P_a \qquad R_{ef} = P_{ef}^*/P^0 \qquad R_z = P_z^*/P_z$$

(4.61e)

Moreover, the isotope ratio of the vapor at the evaporation front, R_{ef}, is arguably equal to $R_{w,ef}/\alpha_{eq}$, where $R_{w,ef}$ is the same as $R_{w,sfc}$ calculated from equation 4.60c. These relationships may be used to show that

$$\frac{R_z - R_a}{R_{ef} - R_a} = \frac{z}{h_z z_{ef}}$$

(4.62)

Interestingly, even though the humidity and the vapor pressure of any isotopic species vary linearly with depth, the variation of the isotope ratio with depth is curvilinear. It is not obvious how to relate these vapor ratios to the isotope ratios of any soil water that might coexist at any given depth, except at the evaporation front.

Barnes and Allison (1983) provide further arguments that the effective depth is directly proportional to $(1 - h_a)$, and inversely proportional to the evaporation rate E, so that

$$z_{ef} = (1 - h_a) \frac{ND_v}{\rho E}$$

(4.63)

where D_v is the effective diffusion coefficient of the vapor which depends on the tortuosity and the degree of saturation of the soil pores, N is the saturated concentration of water vapor in the soil atmosphere, and ρ is the density of water.

Below the evaporation front, at steady state, the expected profile will be basically exponential, similar to that calculated for saturated soils. The result probably takes the following form:

$$\frac{R_z - R_{ef}}{R_{res} - R_{ef}} = \frac{1 - e^{-E(z - z_{ef})/D}}{1 - e^{-E(L - z_e)/D}}$$

(4.64)

However, because isotope transport can actually occur in both the vapor phase and the liquid phase, the diffusion coefficient D may have a complicated form. Barnes and Allison (1983, 1988) provide much additional discussion.

4.6 Kinetic Fractionation Factors

Relations for the binary diffusion coefficient for trace gases in an abundant background may be derived from the kinetic theory of gases. The ratio D/D^* of diffusivities of the isotopomers of the minor gas that has molecular weights of M and M^*, present in the background gas with molecular weight G, simplify to the following relationship (Craig and Gordon, 1965; Kaye, 1987):

$$\frac{D}{D^*} = \sqrt{\frac{M^*}{M} \frac{M+G}{M^*+G}} \qquad (4.65)$$

For water vapor dissolved in ordinary dry air, which has a molecular weight G of 28.966, the ratios of the diffusivities become

$$\frac{D_{H_2O}}{D_{HDO}} = 1.0166 \quad \text{and} \quad \frac{D_{H_2O^{16}}}{D_{H_2O^{18}}} = 1.0324 \qquad (4.66)$$

These values allow for the prediction of the values for α_{evap}^0 for the evaporation of water, using equation 4.56a, b. However, considerable disagreement exists regarding the appropriate power (n) for the D/D^* term in these predictions. Simple diffusion theory predicts a value of 0.5 for evaporation from a plane sheet (equation 4.56a, b), or from spheres for short times, to a value of 1.0 for evaporation from spheres for long times. Other proposals include a value of 0.58 for experiments with water drops (Stewart, 1975), a value of 0.67 for laminar flow situations, and a value of 1.0 for the possible situation in soils. That is,

$$\alpha_{evap}^0 = \alpha_{eq} \left(\frac{D}{D^*}\right)^n \qquad (4.67)$$

While it is not possible to resolve these matters here, it is straightforward to present a table of possible values (Table 4.2), using water–water vapor equilibrium fractionation factors in Table 3.2. These values compare reasonably well with experiment. For example, for drops evaporating into a dry nitrogen atmosphere at 7°C, Stewart (1975; expt. A) measured values of 1.114 and 1.0288 for the hydrogen and oxygen α_{evap}^0 factors, respectively. It is possible that different factors, corresponding to a power n of 1.0, are appropriate for evaporation from soils, because the soil water would arguably behave much more like small disseminated drops than like a plane sheet. On graphs of δD versus $\delta^{18}O$, the

Table 4.2 Calculated Values for α_{evap}^0 for Hydrogen and Oxygen (Equation 4.67)

Power (n)	$\alpha_{evap,D}^0, \alpha_{evap,^{18}O}^0$	$\alpha_{evap,D}^0, \alpha_{evap,^{18}O}^0$	$\alpha_{evap,D}^0, \alpha_{evap,^{18}O}^0$	$\alpha_{evap,D}^0, \alpha_{evap,^{18}O}^0$
	0°C	10°C	20°C	100°C
0.50	1.122, 1.0280	1.107, 1.0270	1.094, 1.0261	1.0355, 1.0212
0.58	1.123, 1.0307	1.108, 1.0296	1.095, 1.0287	1.0369, 1.0238
0.67	1.127, 1.0336	1.110, 1.0326	1.097, 1.0317	1.0385, 1.0268
1.00	1.131, 1.0446	1.116, 1.0435	1.103, 1.0426	1.0441, 1.0376

evaporation trends for soils indeed have lower slopes than are observed for standing bodies of water.

One last complication is that the transport mechanics become more complex when the effects of wind and turbulence are involved. One possibility suggested by the work of Merlivat and Jouzel (1979) is that these effects require inclusion of another factor:

$$\alpha_{evap}^0 = \alpha_{eq} \frac{\left(\dfrac{D}{D^*}\right)^n + \dfrac{\rho_T}{\rho_M}}{1 + \dfrac{\rho_T}{\rho_M}} \tag{4.68}$$

where ρ_T/ρ_M is the ratio of the turbulent and molecular resistances, defined by aerodynamic models. However, Merlivat and Jouzel (1979) note that similar results are obtained over a wide range of wind velocities.

4.7 Isotopic Distribution in the Atmosphere

4.7.1 Gravitational Fractionation

Classical thermodynamics predicts that a mixture of gases would partially segregate in an unperturbed, vertical column in a gravity field. The condition for hydrostatic equilibrium is

$$dP/dz = -\rho g \tag{4.69}$$

where P is pressure, z is vertical distance, ρ is the density of the material, and g is gravitational acceleration. For an incompressible fluid, density is constant and this leads to the familiar equation of hydrostatics, $P = \rho gz$. For a compressible material, such as a gas, account must be made of the variations of density with pressure, which may be accomplished with the ideal gas law. Thus,

$$d \ln P/dz = -gM/RT \tag{4.70a}$$

where the density was removed by introducing the molecular weight M of the gas. For isothermal conditions, and using the boundary condition that the reference pressure is P_0 at depth zero,

$$\ln \frac{P}{P_0} = -\frac{gzM}{RT} \tag{4.70b}$$

For a different gas or isotopomer that has molecular weight M^*, a similar equation may be written:

$$\ln \frac{P^*}{P_0^*} = -\frac{gzM^*}{RT} \tag{4.70c}$$

Now, subtracting 4.70b from equation 4.70c gives

$$\boxed{\ln \frac{R}{R_0} = -\frac{(M^* - M)gz}{RT}} \tag{4.71a}$$

where R is the isotope ratio, or, alternatively, the molar ratio of two different gases, and R_0 is the reference value. Of course, this result may be directly rewritten in terms of δ-values:

$$\ln \frac{1000 + \delta}{1000 + \delta_0} = -\frac{(M^* - M)gz}{RT} \qquad (4.71b)$$

Several different forms and approximations of the last two equations may be found in the literature. Kaye (1987) provides expressions, with production and loss terms, for the steady-state gravitational distribution of gases.

Earth's lower atmosphere is much too turbulent for gravitational fractionation to be observed. While it is possible that certain gases in the upper atmosphere could be affected by this process, compelling data for this have not been found (e.g., see Figure 4.15). However, air trapped in deep snow packs can develop such profiles, by slow diffusion. This process can change the ratios of the constituent gases—for example, the Ar/N_2 ratio—as well as fractionate the oxygen and nitrogen isotopes of the air molecules (e.g., Sowers et al., 1992).

4.7.2 Loss to Space

The elementary kinetic theory of gases indicates that, at a given temperature, the kinetic energy of an ideal gas depends only on the temperature, being equal to the quantity $3RT/2$. Equating this result with the classical kinetic energy gives the value for the root-mean-square velocity (V_{rms}) of the molecules:

$$V_{rms} = \sqrt{\frac{3RT}{M}} \qquad (4.72)$$

where M is the molecular weight of the gas.

A far more remarkable description of gas velocities is given by the Maxwell-Boltzmann distribution law. This distribution defines the probability distribution p for molecular velocities, expressed as a function of the temperature, molecular weight M, and velocity V:

$$p = 4\pi V^2 \left(\frac{M}{2\pi RT}\right)^{3/2} e^{-MV^2/(2RT)} \qquad (4.73)$$

Figure 4.14 shows the actual distribution of velocities of H_2 and HD molecules at 700 K. While these molecules have the same average kinetic energy, it is clearly seen that the H_2 molecules have systematically higher velocities. It may also be shown that the value for V_{rms} calculated from the Maxwell-Boltzmann distribution agrees perfectly with the result from the kinetic theory of gases. In addition, the most probable velocity, V_{mp}, indicated by the maximum of the distribution curve, may be found by finding $\partial p / \partial V$ and setting the result to zero:

$$V_{mp} = \sqrt{\frac{2RT}{M}} \qquad (4.74)$$

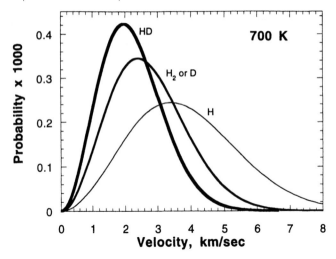

Figure 4.14 Distribution of velocities of gaseous H_2 and HD molecules and monatomic H and D at 700 K, calculated from the Maxwell-Boltzmann equation (4.73). The velocities of the HD molecules are systematically lower than the practically identical velocities of D and H_2, and they, in turn, are systematically lower than those of H. While the areas under all the curves are identical, the peaks of the various curves represent the most probable velocities which increase with decreasing molecular weight. The proportion of molecules that have velocities that approach Earth's escape velocity of 11.2 km/s is small for all of these species, but, given sufficient time, appreciable H could be preferentially lost.

For many situations that involve gases, the associated isotopic fractionation effects depend on the ratios of the velocities of the various isotopomers. Thus, the ratio of either the rms velocities or the most probable velocities for two gases or isotopomers is

$$\frac{V}{V^*} = \sqrt{\frac{M^*}{M}} \qquad (4.75)$$

For certain situations, this ratio will be identical to the ratio of the diffusivities, giving "Graham's law" of diffusion, as discussed earlier. Of more importance here are the high-velocity "tails" of the distribution function. Only the fraction of the molecules that have these high velocities will have sufficient velocity to escape a planet. The fraction of molecules that have velocities that exceed a given value may be determined by integrating the Maxwell-Boltzmann distribution (R. E. C., unpublished):

$$\Phi_{V \geq V_r} = \int_{V_r}^{\infty} p \, dV = \sqrt{\frac{2M}{\pi RT}} V_r e^{-MV_r^2/(2RT)} + \left(1 - \text{erf}\left[\sqrt{\frac{2M}{\pi RT}} V_r\right]\right) \qquad (4.76)$$

Note that if the reference velocity V_r is taken as zero, then the fraction is unity, corresponding to the obvious requirement that all of the molecules must have a finite velocity.

The isotopic fractionation factor for the loss of high-velocity molecules will be equal to the ratio of such fractions for the common and trace isotopes. That is,

$$\alpha_{\text{loss}} = \frac{\Phi^*_{V^* \geq V_r}}{\Phi_{V \geq V_r}} \qquad (4.77)$$

For any given problem, this fractionation may be found from the previous two expressions. However, for high reference velocities, the term in parenthesis in equation 4.76 is extremely small, so that, to a good approximation,

$$\alpha_{\text{loss}} \cong \sqrt{\frac{M^*}{M}} e^{-(M^*-M)V_r^2/(2\mathbf{R}T)} \qquad (4.78)$$

In order to apply this expression to loss from planetary atmospheres, the reference velocity needs to be determined. Arguably, the velocity of interest is closely related to the escape velocity. Neglecting rotational effects, the escape velocity may be determined by simply equating the classical kinetic energy of an object with its gravitational potential energy on a planet with mass M_p and radius r. The well-known result is independent of the particle mass:

$$V_{\text{esc}} = \sqrt{\frac{2GM_p}{r}} = \sqrt{2gr} \qquad (4.79)$$

where G is the universal gravitational constant and g is the gravitational acceleration for the particular planet. For Earth, the escape velocity is close to 11.2 km/s, while it is a staggering 59.4 km/s for Jupiter.

Given the high escape velocities for major planets, the values for α_{loss} are mostly very close to zero. In other words, the light isotope is lost almost exclusively relative to the heavy isotope. The involvement of molecular dissociation reactions will complicate the analysis, but generally will only enhance the preferential loss of the light isotope. As a result, in most cases, the Rayleigh fractionation equation for a trace isotope reduces to

$$\frac{R}{R_i} = \frac{1}{f} \qquad (4.80)$$

where f is the fraction of the gas of interest that remains after the amount $1 - f$ has been lost to space.

4.8 Mass-Independent Fractionation

In practically all physiochemical processes, any isotope fractionation effects are mass dependent, and are governed by the mechanisms that were originally studied by Urey (1947). For light elements composed of three or more stable isotopes, the fractionation factors for any pair will increase with the mass difference between the isotopes, as codified by the new equations 2.74 and 2.73b presented above. However, a relatively new and unexpected effect, called "mass-independent" fractionation, produces large isotope fractionations that are independent of the mass differences between the various pairs of isotopes.

Oxygen isotopes provide the best evidence for behavior of this latter type. Normally, for practically any material, the $^{17}O/^{16}O$ fractionation factor (α^{17}) will be close to the square root of the $^{18}O/^{16}O$ fractionation factor (α^{18}), in conformity to equation 2.74. That is, for mass-dependent fractionation,

$$\alpha^{17} = (\alpha^{18})^{0.529} \tag{4.81}$$

On a graph where $\delta^{17}O$ values are plotted against the $\delta^{18}O$ values for the same material, the result of any "normal" process will be a nearly straight line with a slope close to 1/2. However, for "mass-independent" processes, the $^{17}O/^{16}O$ fractionations are very unusual in that they are essentially the same as the $^{18}O/^{16}O$ fractionations; that is,

$$\alpha^{17} \cong \alpha^{18} \tag{4.82}$$

Mass-independent fractionations were first found in rare refractory inclusions in certain meteorites (see chapter 5). Subsequently, these effects have been found in certain gases in Earth's stratosphere and mesosphere, including ozone, carbon dioxide, and carbon monoxide (e.g., Thiemens and Heidenreich, 1983; Thiemens et al., 1995). Such effects for CO_2 are shown in Figure 4.15; were the $\delta^{17}O$ values plotted directly against those of $\delta^{18}O$, a straight line with a unit slope would be

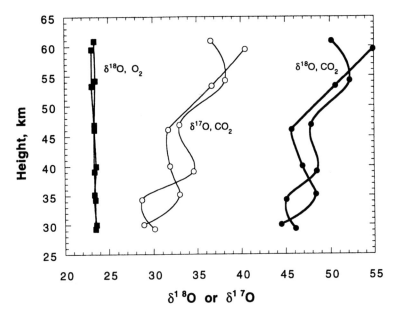

Figure 4.15 Vertical variations of the $\delta^{18}O$ values of gaseous CO_2 and O_2, and of the $\delta^{17}O$ values of CO_2, in the stratosphere and mesosphere, collected on two different rocket flights. The variations in the CO_2 are unusual, and exhibit the peculiar character of mass-independent fraction, such that the $\delta^{17}O$ variations have the same magnitude as those of $\delta^{18}O$. The $\delta^{18}O$ values of O_2 may decrease slightly with height. Data from Thiemens et al. (1995).

obtained. The aforenamed gases participate in molecular reactions that involve ultraviolet photodissociation or spark discharge, and, in some way, the molecular selection rules of these processes circumvent the normal effects of isotopic fractionation.

Unless otherwise stated, mass-independent fractionation will not be involved in the processes discussed in this book. Nevertheless, this phenomenon is one of the most noteworthy isotope effects discovered in recent years, and further study of it may lead to a more encompassing theory of isotope fractionation.

4.9 Summary

Isotopic disequilibrium is a common phenomenon. In many cases, the distribution of isotopes under nonequilibrium conditions may be evaluated with the laws of elementary kinetics. For homogeneous systems, and for systems where spatial gradients in isotope concentrations are not of primary interest, this theory provides a powerful yet simple viewpoint. The differential equations in this chapter may be applied to many different chemical and physical situations.

For other systems, and particularly for systems where spatial isotopic gradients are important, the laws of diffusion may be applied with advantage. In many cases, the equations are mathematically complex, but numerous solutions are provided in standard books. Moreover, the equations are simplified if steady-state conditions apply. In simple systems, almost perfect agreement may be secured between the predictions of diffusion theory and kinetic theory for isotopic systems.

4.10 Problems

1. A leaky tank contains chlorine originally having a normal isotopic ratio, R_N. Presuming that the gas escape is governed by "Graham's law" of diffusion, what is the effective isotopic fractionation factor for the loss? If, at some later time, the $^{37}Cl/^{35}Cl$ ratio equals 1.04 R_N, what fraction of the gas has been lost?

2. In an experiment at 19.5°C, 1 atm of CO_2 gas with $\delta^{18}O = +3.36‰$ is placed in a closed, 1-liter vessel that contains 100 g of water with $\delta^{18}O = -8.25‰$. After only 31 min of shaking the vessel, the $\delta^{18}O$ value of the gas is found to be $+25.78‰$. At what time will the $\delta^{18}O$ value of the gas be equal to $+30.0‰$? Calculate the final equilibrium $\delta^{18}O$ value of the CO_2. Would the latter result be different if the 1-liter vessel contained only 5 g of water, and, if so, by what amount?

$$\text{Note: } 1000 \ln \alpha_{CO_2-H_2O} = 16.60 \times 10^3/T - 15.19$$

Range: 0–100°C (Friedman and O'Neil, 1977)

3. A very long aqueduct, tapping a source of mountain water with a $\delta^{18}O$ value of -10, flows through a semiarid region with an average atmospheric humidity of

35%. If the atmospheric vapor typically has a $\delta^{18}O$ value of -18, and if the water at the aqueduct terminus has a value of -8.5, what fraction of the water was evaporated en route?

4. A terminal lake in an arid region receives inflow with δD and $\delta^{18}O$ values of -86 and -12, respectively, and evaporates into an atmosphere with a very low average humidity of 25% that contains water vapor with δD and $\delta^{18}O$ values of -150 and -20, respectively. Determine (1) the δD and $\delta^{18}O$ values of the lake at steady state, (2) the δD and $\delta^{18}O$ values of the isotopic end state that would be attained by the last drop of water in an evaporation pan at this location, and (3) the salinity of the lake.

5. A shallow, flow-through lake in a temperate region receives inflow with δD and $\delta^{18}O$ values of -46 and -7, respectively, and evaporates into an atmosphere with an average humidity of 80% that contains water vapor with δD and $\delta^{18}O$ values of -95 and -15, respectively. Determine the δD and $\delta^{18}O$ values of the lake at steady state, assuming that 10% of the inflowing water is lost to evaporation. Also, plot a graph of the δD and $\delta^{18}O$ values versus f that would be observed in two evaporation pans at this location that were initially charged with water that had δD and $\delta^{18}O$ values of -110 and -15 in the first case, and an enriched water with $+40$ and $+10$ in the second case. Also, determine the slopes that these pans exhibit on a graph of δD versus $\delta^{18}O$.

6. Upward flow of deep groundwater with δD and $\delta^{18}O$ values of -30 and -5, respectively, maintains saturated soil in a desert oasis. In this area, the average humidity of 20% represents advected marine vapor with δD and $\delta^{18}O$ values of -100 and -17, respectively. In the following, assume that the self-diffusion constant of water in the soil is approximately 10^{-5} cm^2/s, and that the evaporation rate is 3 cm/day. (a) Determine the δD and $\delta^{18}O$ values of a small, shallow pool of water at the oasis. (b) Calculate and plot a graph of the vertical profiles of the δD and $\delta^{18}O$ values of soil water. (c) Plot a graph of the δD versus the $\delta^{18}O$ values, and determine the slope of the trend.

7. The O_2 gas at a height of 30 km in Earth's atmosphere has a $\delta^{18}O$ value of $+23.5$. Predict the value the gas would have at a height of 60 km if gravitational fractionation exerted the primary control on the isotope distribution. Compare your result with the data in Figure 4.15, which suggest a very weak, ca. 0.01‰ per km, reduction in the $\delta^{18}O$ values of O_2 with height.

8. Integrate the Maxwell–Boltzmann distribution law to determine a relationship for the average velocity of gas molecules; that is, determine

$$\bar{V} = \int_0^\infty pc\, dc$$

Why does this result differ from the root-mean-square velocity indicated by the elementary kinetic theory of gases?

9. The D/H ratio of Earth is approximately $5\times$ greater than the Jovian ratio. Assuming that this is the result of the preferential loss of protium to space over geologic time, what fraction of Earth's hydrosphere has been lost?

10. Titan, the only moon in the solar system with a significant atmosphere, has an escape velocity of 2.6 km/s. The temperature is only about 92 K on Titan's surface and 160K in its upper atmosphere. Which of the gases H_2, HD, CH_4, N_2, CO_2, and Ar would be expected to be retained in Titan's atmosphere? Which of these gases might be expected to show significant enrichment in the heavy isotopes?

11. According to Lide (1991), the diffusion coefficient of water vapor in air is 0.239 cm^2/s. What values would be predicted for HDO and $H_2{}^{18}O$, according to (a) Graham's law of diffusion and (b) equation 4.65?

12. Show that equation 4.53c is the correct solution to the boundary value problem set up in section 4.5.2. To do this, show that (1) the solution fits the initial and boundary conditions specified in equations 4.53a and b, and (2) show by differentiation that the solution satisfies Fick's second law for one-dimensional transport (equation 4.48b). Several relationships in Appendix A.3 will help.

References

Barnes, C.J. and Allison, G.B. (1983) The distribution of deuterium and ^{18}O in dry soils. Theory. *J. Hydrol.*, **60**, 141–156.
Barnes, C.J. and Allison, G.B. (1988) Tracing of water movement in the unsaturated zone using stable isotopes of hydrogen and oxygen. *J. Hydrol.*, **100**, 143–176.
Carslaw, H.S. and Jaeger, J.C. (1976) *Conduction of Heat in Solids.* Oxford University Press, Oxford.
Cole, D.R. and Ohmoto, H. (1986) Kinetics of isotopic exchange at elevated temperatures and pressures. *Rev. Mineral.*, **16**, 41–90.
Craig, H. and Gordon, L.I. (1965) Deuterium and oxygen 18 variations in the ocean and the marine atmosphere. Symposium on Marine Geochemistry, Narraganset Marine Laboratory, University of Rhode Island Publication, Vol. 3, pp. 277–374.
Craig, H., Gordon, L.J., and Horibe, Y. (1963) Isotopic exchange effects in the evaporation of water. *J. Geophys. Res.*, **68**, 5079–5087.
Crank, J. (1975) *The Mathematics of Diffusion.* Oxford University Press, Oxford.
Criss, R.E., Gregory, R.T., and Taylor, H.P., Jr. (1987) Kinetic theory of oxygen isotopic exchange between minerals and water. *Geochim. Cosmochim. Acta*, **51**, 1099–1108.
Ehhalt, D. and Knott, K. (1965) Kinetische Isotopentrennung bei der Verdampfung von Wasser. *Tellus*, **17**, 389–397.
Epstein, S. and Mayeda, T.K. (1953) Variations of O^{18} content of waters from natural sources, *Geochim. Cosmochim. Acta*, **4**, 213–224.
Fortier, S.M. (1994) An on-line experimental/analytical method for measuring the kinetics of oxygen isotope exchange between CO_2 and saline/hypersaline salt solutions at low (25–50°C) temperature. *Chem. Geol.*, **116**, 155–162.
Friedman, I. and O'Neil, J.R. (1977) Compilation of stable isotope fractionation factors of geochemical interest. In *Data of Geochemistry*, Fleischer, M., ed. U.S. Geological Survey Professional Paper 440 KK.

Gonfiantini, R. (1986) Environmental isotopes in lake studies. In *Handbook of Environmental Isotope Geochemistry*, Vol. 2, Fritz, P. and Fontes, J.Ch., eds. Elsevier, New York, pp. 113–168.

Gregory, R.T., Criss, R.E., and Taylor, H.P., Jr. (1989) Oxygen isotope exchange kinetics of mineral pairs in closed and open systems: applications to problems of hydrothermal alteration of igneous rocks and Precambrian iron formations. *Chem. Geol.*, **75**, 1–42.

Ingraham, N.L. and Criss, R.E. (1993) The effects of surface area and volume on the rate of isotopic exchange between water and vapor, *J. Geophys. Res.*, **98**, 20547–20553.

Ingraham, N.L. and Criss, R.E. (1998) The effect of vapor pressure on the rate of isotopic exchange between water and vapor. *Chem. Geol.*, **150**, 287–292.

Ingraham, N.L., Criss, R.E., and Rose, T.P. (1991) The effects of kinetic fractionation and isotopic exchange during the evaporation of water. *Trans. Am. Geophys. Union*, **72**, 178.

Kaye, J.A. (1987) Mechanisms and observations for isotope fractionation of molecular species in planetary atmospheres. *Rev. Geophys.*, **25**, 1609–1658.

Lide, D.R. (1991) *Handbook of Chemistry and Physics*, 71st ed. CRC Press, Inc., Boston.

Merlivat, L. and Jouzel, J. (1979) Global climatic interpretation of the deuterium–oxygen 18 relationship for precipitation. *J. Geophys. Res.*, **84**, 5029–5033.

O'Neil, J.R. and Taylor, H.P., Jr. (1967) The oxygen isotope and cation exchange chemistry of feldspar. *Am. Mineral.*, **52**, 1414–1437.

Sowers, T., Bender, M., Raynard, D., and Korotkevich, Y.S. (1992) $\delta^{15}N$ of N_2 in air trapped in Polar ice: a tracer of gas transport in the firn and a possible constraint on ice age–gas age differences. *J. Geophys. Res.*, **97**, 15683–15697.

Stewart, M.K. (1975) Stable isotope fractionation due to evaporation and isotopic exchange of falling waterdrops: applications to atmospheric processes and evaporation of lakes. *J. Geophys. Res.*, **80**, 1133–1146.

Stewart, M.K. and Friedman, I. (1975) Deuterium fractionation between aqueous salt solutions and water vapor. *J. Geophys Res.*, **80**, 3812–3818.

Thiemens, M.H. and Heidenreich, J.E. (1983) The mass-independent fractionation of oxygen: a novel isotope effect and its possible cosmological implications. *Science*, **219**, 1073–1075.

Thiemens, M.H., Jackson, T., Zipf, E.C., Erdman, P.W., and van Egmond, C. (1995) Carbon dioxide and oxygen isotope anomalies in the mesosphere and stratosphere. *Science*, **270**, 969–972.

Urey, H.C. (1947) The thermodynamic properties of isotopic substances. *J. Chem Soc.* (London), 562–581.

Zimmermann, U., Ehhalt, D., and Munnich, K.O. (1967) Soil water movement and evapotranspiration: changes in the isotopic composition of the water. In *Proceedings of the Symposium on Isotopes in Hydrology* (Vienna, 1966). International Atomic Energy Agency (I.A.E.A), Vienna, pp. 567–584.

5

Igneous Rocks, Meteorites, and Fluid–Rock Interactions

5.1 Oxygen and Hydrogen Isotope Geochemistry of Rocks

Oxygen is the most important element in common, rock-forming minerals. Earth's crust and mantle contain about 44 wt. % oxygen, and even with its dense iron core, the bulk Earth is estimated to be approximately 30% oxygen. Considering the low mass and large size of the oxygen atom, the oxygen content is even higher if expressed in terms of vol. % or mol. %.

For the above reason, a central problem of stable isotope geochemistry is to explain the distribution of oxygen isotope ratios in rocks. As shown in this chapter, much of the diversity in the abundance ratios is related to interactions of rocks with Earth's extensive hydrosphere, which is nearly 89 wt. % oxygen. Even though hydrogen is only a minor element in rocks, some discussion of hydrogen isotopes is included here because they provide powerful complementary relationships to evaluate fluid–rock interactions.

5.1.1 Oxygen Isotope Geochemistry

5.1.1.1 Bulk (Primordial) Composition   It has been suspected for centuries, and has now been proven by oxygen isotope data (see later), that Earth and the Moon have very similar origins. In particular, the $\delta^{18}O$ values of large rock reservoirs on the Moon and Earth are practically identical. Diverse lunar lithologies have remarkably uniform values ranging only from +5.4 to +6.8 relative to SMOW, with the subset of lunar igneous rocks showing even less variation at $+5.7 \pm 0.2$ (Epstein and Taylor, 1971; Taylor and Epstein, 1973).

The same limited range of values is found for the largest lithologic reservoirs on Earth (Figure 5.1). For example, mid-ocean ridge (MOR) basalts are the most abundant igneous rock type on Earth, and cover practically the entire ocean floor. The $\delta^{18}O$ values of these basalts are practically uniform at $+5.7 \pm 0.5$ (Kyser, 1986). Similarly, other mafic lavas, as well as peridotites, pyroxenites, and practically all other mantle materials with the exception of the ophiolites and eclogites, have $\delta^{18}O$ values in the restricted range of $+5.0$ to $+8.0$. Moreover, no apparent secular trend over geologic time has been found in the bulk $\delta^{18}O$ values of these reservoirs. For these reasons, it is likely that the bulk $\delta^{18}O$ values of Earth and the Moon are identical and very close to $+5.7 \pm 0.2$. This value therefore provides the appropriate reference point for discussion of the oxygen isotope ratios of terrestrial rocks.

5.1.1.2 Terrestrial Igneous Rocks   In contrast to the uniformity of the mantle reservoir, the $\delta^{18}O$ values of igneous rocks collected throughout the world vary widely—at least from -10.5 to $+16$ per mil (Figure 5.1). Similarly, compared with the Moon, the total range of $\delta^{18}O$ values observed in terrestrial *magmas* is wide— at least $+2$ to $+16$ per mil (Taylor and Sheppard, 1986). As will be discussed in this chapter, much of the variation in rocks is the result of secondary alteration processes, especially fluid–rock interactions. Fortunately, such effects can com-

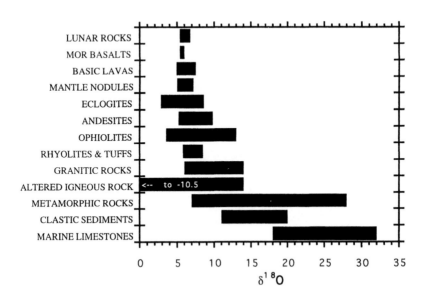

Figure 5.1   Typical ranges of the whole-rock $\delta^{18}O$ values of rocks. The large mantle reservoir on Earth, typified by mid-ocean ridge (MOR) basalts and mantle xenoliths, has a narrow range of values centered near $+5.7$ that is similar to lunar rocks. Other rock types have widely ranging values, generally indicating that they have resided, or contain a component that once resided, near Earth's surface. Modified after Criss (1995), utilizing various data sources.

monly be detected because they create unusual fractionation patterns among the minerals; in addition, hydrothermal alteration processes typically produce anomalous isotope ratios for both hydrogen and oxygen. However, even certain mantle materials, such as eclogites or individual samples from ophiolite sequences, have $\delta^{18}O$ values that vary over a considerable range.

Excepting these complications, the ^{18}O contents of unaltered igneous rocks tend to increase with increasing SiO_2 contents. Thus, the values for volcanic rocks generally increase in the order of basalts to andesites to rhyolites, consistent with their increasing silica concentrations of about 50, 60, and 70 wt. %, respectively. Similarly, for plutonic rocks over this compositional range, the $\delta^{18}O$ values trend to increase in the order of ultramafic rocks, to gabbros, tonalites, granodiorites, and ultimately to granites (e.g., Taylor, 1968).

5.1.1.3 Sedimentary Rocks   The $\delta^{18}O$ values of most sedimentary rocks are higher than those of igneous rocks (Figure 5.1). The deposition and, in many cases, the formation of the minerals in sedimentary rocks occurs in surficial environments that are in intimate contact with the hydrosphere. Fractionation factors between the minerals and water are large in these low-temperature environments. Consequently, the minerals tend to acquire a high ^{18}O character, particularly if they are chemical sediments or weathering products. Even if they are relict minerals from igneous terranes, they will tend to increase in ^{18}O if they undergo any reaction or exchange under surface conditions.

Thus, the $\delta^{18}O$ values of clastic sediments are almost invariably greater than or equal to those of their protoliths. For example, the $\delta^{18}O$ values of beach sands are about +9 to +12.5, and principally reflect inherited igneous and metamorphic quartz (Savin and Epstein, 1970c). Sandstones tend to have slightly higher values, typically $+13 \pm 2.5$; while they also are predominantly composed of inherited quartz from crystalline rocks, the presence of authigenic quartz or calcite cement commonly has increased their whole-rock $\delta^{18}O$ values. Typical $\delta^{18}O$ values of shales are +14 to +19, and reflect the inheritance of protolith materials, as well as the presence of high-^{18}O authigenic clay minerals (Savin and Epstein, 1970b). The $\delta^{18}O$ values of greywackes at $+13 \pm 1.5$ are similar to those for sandstones, and are substantially higher than those of volcanic rocks, probably because of the formation of authigenic minerals, including clays.

Moreover, and because seawater has a higher ^{18}O content than most freshwaters, on Earth the rocks with the highest $\delta^{18}O$ values are close to being in isotopic equilibrium with seawater. Many such rocks have precipitated directly from seawater, or have otherwise approximately equilibrated with seawater. The most important example, recent limestones, are approximately 30‰ richer in ^{18}O than the seawater in which they form. Similarly, cherts are as much as 36‰ richer than the seawater from which they precipitate (Knauth and Epstein, 1976). These high values are a direct consequence of the large calcite–water and quartz–water fractionations that occur at the low temperatures of Earth's surface (see Appendix A.4).

The $\delta^{18}O$ values of older chemical sediments such as ancient limestones, cherts, or banded iron formations, are also quite high (mostly $> +18$), but the values generally decrease with increasing age. One possible explanation for this difference is that surface temperatures were substantially higher in ancient times. Alternatively, a large shift in the $\delta^{18}O$ value of seawater may have occurred over geologic time, but this is not widely accepted (see later). In all likelihood, the principal reason is that secondary alteration processes, which mostly occur in subsurface environments where temperatures are elevated and where pore fluids are modified, have systematically reduced the $\delta^{18}O$ values (e.g., Keith and Weber, 1964; Anderson and Arthur, 1983).

5.1.1.4 Metamorphic Rocks Metamorphic rocks have highly variable $\delta^{18}O$ values that tend to be intermediate between those of igneous rocks and recent sediments. Orthoquartzites, granitic gneisses, and marbles may inherit the $\delta^{18}O$ values of their protolith materials without substantial change. More commonly, the $\delta^{18}O$ values of rocks become progressively reduced with increasing grades of metamorphism. In particular, in numerous regional metamorphic terranes, the $\delta^{18}O$ values of pelitic metasediments have been reduced from their original sedimentary values (e.g., Garlick and Epstein, 1967). Similarly, the $\delta^{18}O$ values of carbonates are progressively lowered, by as much as 20‰, with increasing grade or decreasing distance to igneous contacts (Valley, 1986). In addition, in several high-grade areas, such as in large amphibolite zones in southern Ontario, New York, and the Idaho panhandle, the $\delta^{18}O$ values of diverse rock types have been "homogenized" to a common, igneous-like value by pervasive, large-scale exchange that involves metamorphic fluids (Taylor, 1970; Shieh and Schwarcz, 1974; Criss and Fleck, 1990). In contrast, the $\delta^{18}O$ values of different rock types remain heterogeneous in certain other areas. In many such zones, small-scale, diffusion-like ^{18}O gradients may be produced by isotopic exchange between proximal contrasting rock units, particularly between impermeable carbonates and interbedded pelites (Rye et al., 1976; Nabelek et al., 1984).

5.1.2 Hydrogen Isotope Geochemistry

5.1.2.1 Bulk (Primordial) Composition The bulk isotopic composition of hydrogen is more difficult to establish than that for oxygen. Hydrogen is only a minor element in common rocks and minerals, and its isotopic variations are large in solar system materials. A truly enormous range of values, -500 to $+9000$ per mil, is found in meteorites. The Moon provides no special data for comparison, because it experienced nearly complete volatile loss early in its history, and because it is constantly bombarded by the solar wind, which is practically devoid of deuterium. Most hydrogen on Earth resides in the hydrosphere, which has a high δD value, near -10, because it is dominated by seawater. Because most terrestrial rocks have δD values between -40 and -95, the bulk δD value of Earth is probably only slightly lower than the hydrosphere, perhaps -20‰. This value is significantly higher than for most other planets in the solar system, probably due to the progressive loss of protium to space over geologic time (see later).

5.1.2.2 Hydrogen Isotopes of Terrestrial Rocks   While the δD values of meteorites and lunar rocks vary enormously, the δD values of virtually all *unaltered* igneous, metamorphic, and sedimentary rocks on Earth are in the comparatively narrow range of -40 to -95 (Taylor and Sheppard, 1986). Magmatic and deepseated waters on Earth also fall in this range (Sheppard and Epstein, 1970). The δD values of hydrous phases are typically 30 ± 20 per mil lower than those of the water that would be in isotopic equilibrium at temperatures where they are stable (e.g., Suzuoki and Epstein, 1976). Thus, the δD values observed for terrestrial rocks are similar to those of clay and alteration minerals formed in contact with seawater or with the abundant meteoric waters of tropical and temperate zones. In fact, on a plot of δD versus $\delta^{18}O$, many clays fall along lines that parallel the MWL, being offset by the ^{18}O and D fractionation factors between the particular clay mineral and water (Savin and Epstein, 1970a).

The similarity of the δD values for most rocks on Earth, whether they be igneous, metamorphic, or sedimentary, suggests that hydrogen is being continuously recycled through Earth's upper zones, via the process of alteration (hydration), subduction, and dehydration (Taylor, 1986). Thus, the H_2O in magmatic and deep-seated rocks appears to be mostly derived from the dehydration of surface materials, such as clays that have been subducted at trenches (Taylor and Sheppard, 1986). Over geologic time, subducted water completely dominates the δD value of the upper mantle, so that a "steady-state" relationship has been attained. This process effectively buffers the δD values for primary rocks and deep-seated waters to be in this -40 to $-95‰$ range. Superimposed on this range are larger variations produced by secondary alteration processes, especially hydrothermal disturbances (see later).

5.2 Igneous Rocks: Primary Crystallization

5.2.1 Equilibrium Fractionations

To explain the diversity of oxygen isotope ratios on Earth, it is necessary to first establish the consequences of magmatic processes on the oxygen isotope ratios of minerals. Once these magmatic patterns are documented, the effects produced by subsolidus alteration processes can be distinguished. An invaluable criterion is provided by the ^{18}O fractionation patterns of minerals that are produced during crystallization from an igneous melt. These fractionations are small as a consequence of the high temperatures of magmas. Although some complications arise from compositional variations, measurements on rocks indicate a strong tendency for the ^{18}O contents of igneous minerals to decrease in the following order:

Quartz > Alkalic (K Na) feldspars > Calcic feldspars > Muscovite >

Pyroxenes and amphiboles > Biotite > Olivine > Ilmenite > Magnetite

$$(5.1)$$

It is useful to compare this observed relationship with experimentally determined fractionation factors (e.g., Table 2.3 and Appendix A.4; Friedman and O'Neil,

1977; Chiba et al., 1989). This sequence clearly represents a strong tendency for igneous minerals to form in, and thereafter preserve, the pattern of mutual isotopic equilibrium appropriate for magmatic temperatures.

5.2.2 Rayleigh Fractionation

5.2.2.1 Simple Fractional Crystallization   The process of fractional crystallization, enunciated a century ago by G. F. Becker, is thought to play a major role in the production of chemical variations in igneous rocks. In this process, the separation of crystallizing minerals from their parent melt can drive changes in melt chemistry, because the crystals have simple chemical compositions that differ from the bulk composition of complex melts. In natural systems, this process is very complex and depends on many physical and compositional variables. In general, melts undergoing fractional crystallization typically evolve into more siliceous magmas that are progressively enriched in volatiles and in "incompatible" elements that are not easily accommodated into the lattices of simple, rock-forming minerals.

Fractional crystallization can also produce variations in the isotope ratios of light elements, according to the Rayleigh equation discussed in earlier chapters. That is,

$$\left(\frac{R}{R_i}\right)_{melt} = f^{\alpha-1} \tag{5.2}$$

where R refers to the isotope ratio of the melt at any time. All that is required for the production of substantial variations during crystallization is that the crystal–melt fractionation factor

$$\alpha_{xl-melt} = \frac{R_{xl}}{R_{melt}} \tag{5.3}$$

differs significantly from unity. This condition is potentially significant for several of the light fractionating isotopes, such as O, S, and H, but it is thought not to be realized for the isotopes of "heavy" elements, such as Sr, U, Pb, and so on.

The situation therefore hinges on the magnitude of $\alpha_{xl-melt}$. For important rock-forming minerals, such as quartz, feldspar, pyroxenes, olivine, common micas, and amphiboles, the fractionation takes on relatively small values of less than 2‰; that is,

$$0.998 < \alpha_{xl-melt} < 1.002 \tag{5.4a}$$

Perhaps the most important factor is for feldspars:

$$0.9995 < \alpha_{plag-melt} < 1.0006 \tag{5.4b}$$

As pointed out by Taylor and Sheppard (1986), this latter fractionation factor exhibits the strange behavior of being less than unity for sodic feldspars that crystallize from rhyolitic melts, but being greater than unity for the more calcic feldspars that crystallize from basaltic melts. This trend is opposite to the behavior that would be expected from the character of the mineral phase alone (equation 5.1), indicating that the properties of the melt change with composition.

Because the crystal–melt fractionation factors in equations 5.4a and b are small, Rayleigh fractionation that operates during simple fractional crystallization normally cannot produce large variations in the oxygen isotope content of igneous rocks (see Figure 5.3). While the early crystallization of dark minerals such as olivine, magnetite, pyroxene, ilmenite, hornblende, or biotite will tend to increase the $\delta^{18}O$ value of their parent melt, the crystallization of quartz will produce a decrease in the melt $\delta^{18}O$ value. The crystallization of plagioclase, which commonly is the most important mineral, will generally produce only very small changes because the fractionation factor is so small, and because the partitioning of ^{18}O between the crystals and the melt reverses sign with composition. Overall, the fractional crystallization process is not sufficient to cause large variations in the $\delta^{18}O$ values of igneous rocks. This inference conforms with the observation that only a slight (< 1.5‰) ^{18}O enrichment accompanies simple fractional crystallization in well-studied rock suites (see Taylor and Sheppard, 1986).

5.2.2.2 Volatile Separation   While Rayleigh fractionation generally produces only small ^{18}O variations in igneous rock suites, this process can produce very large variations in the D/H ratio of magmas during volatile degassing (Taylor, 1986). For this process, the relevant parameter is the D/H fractionation factor between the vapor and the melt. This parameter has a significant magnitude of about 15–35‰; that is,

$$\alpha_{\text{vapor}-\text{melt}} \sim 1.015 - 1.035 \qquad (5.4c)$$

Rhyolitic magmas, particularly obsidian domes and flows, provide excellent examples of this process (Figure 5.2; Taylor, 1986). The initial stages of eruption of these water-rich mamas are commonly explosive, producing tephra, but progressive volatile degassing occurs so that later eruption stages are more quiescent, and obsidian flows are produced. Because the vapor is enriched in deuterium relative to the melt, the D/H ratio of the remaining magma decreases with time, in a manner that can be modeled with the Rayleigh equation (Figure 5.2). For example, for a degassing melt that originally had a δD value of -50‰, and an initial concentration of dissolved water of 2.0 wt. %, equation 5.2 can be used to readily calculate a δD value of -118.5 for the melt that had only 0.1 wt. % of remaining water. Interestingly, the degassing process produces a different trend than that imprinted by meteoric-hydrothermal alteration (see section 5.4.4), which generally produces the opposite trend of decreasing δD values with increasing water contents.

5.2.3 Assimilation–Fractional Crystallization (AFC)

5.2.3.1 AFC Processes   The $\delta^{18}O$ variations ($+2$ to $+16$‰) observed in igneous magmas are much too large to be the result of simple fractional crystallization. Available data seemingly require interaction of magmas with crustal reservoirs, as these can have comparatively high, and sometimes very low, $\delta^{18}O$ values. For

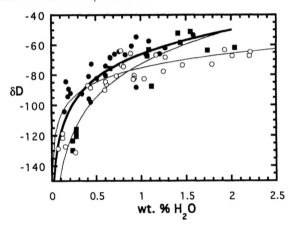

Figure 5.2 Graph of δD values versus the water content of obsidian tephra and flows from Inyo Craters (filled circles), Mammoth Lakes (squares), and the Medicine Lake Highlands (open circles), California. The correlated variations arise from progressive degassing of magmas that were initially rich in water and deuterium, and had explosive (Plinian) character, to comparatively quiescent eruptions of water-poor, D-depleted flows. Three Rayleigh curves are illustrated, representing initial water contents of 2.0, 2.0, and 3.0 wt. %, initial δD values of −50, −50, and −60‰, and vapor–magma D/H fractionation factors of 1.035, 1.025 (heavy curve), and 1.015. Data from Taylor (1986).

example, a straightforward means to form high-^{18}O siliceous magmas is by the incorporation of high-^{18}O sedimentary or metamorphic materials into magmas. Similarly, magmas with anomalously low $\delta^{18}O$ values must have assimilated low-^{18}O materials. As pointed out by H. P. Taylor, it is likely that any rock with a $\delta^{18}O$ value outside the range of +5 to +8 either includes or exchanged with a component that resided near Earth's surface at some time in its history.

The process of assimilating cold country rock into an igneous melt might seem to be a simple, two-endmember mixing process. However, the famous petrologist N. L. Bowen has shown (in 1928) that it is not. At least three endmembers are required, notably the magma, the country rocks, and cumulate minerals. The formation of the latter is a necessary part of the process, because the energy required to heat, melt, and dissolve the country rock must be provided by the magma, either from superheat or from the latent heat of crystallization of minerals that are induced to freeze (see Taylor, 1980). In other words, Bowen clearly realized that the assimilation of cold country rocks into a magma typically promotes the crystallization of cumulate minerals that are already forming. Consequently, assimilation does not radically change the major-element chemistry of the melt, but it does increase the proportions of the late-stage differentiates, and it can drastically alter the concentrations of trace elements and minor isotopes.

5.2.3.2 AFC Differential Equations   The AFC process may be modeled with appropriate calculations and differential equations. The mass of magma at any time, M_m, is affected by the addition of assimilated country rocks at a rate A, and by the subtraction of cumulates at a rate that is equal to the rate A multiplied by the constant r. A symbolic representation of the concentrations and the isotope ratios of an element of interest for the three endmembers is given in Table 5.1.

A few words of explanation are necessary. In Table 5.1, the distribution coefficient D for an element of interest is given by the ratio of concentrations in the cumulate crystals to that in the melt:

$$D = C_{xls}/C_{melt} \qquad (5.5a)$$

This coefficient is identical to the "β" parameter used by Taylor (1980). Not only do the elemental concentrations in these two endmembers differ, but their respective isotope ratios can differ by the fractionation factor α. Moreover, the isotope ratios (R values) used here actually represent the mass of the isotope of interest normalized to the total mass of that element. This definition differs somewhat from the standard "R" used elsewhere in this book, but in cases involving trace isotopes such as ^{18}O, this difference will not be consequential (see chapter 1).

The constant r, representing the mass of cumulates to the mass of rock assimilated, is a key parameter:

$$r = \frac{\text{mass cumulates}}{\text{mass assimilated}} \qquad (5.5b)$$

Note that the volume of magma must necessarily decrease for $r > 1$, and must increase in the less likely case where $r < 1$. Interesting special cases are for r equal to zero, unity, or infinity, which respectively correspond to simple mixing (no cumulates), zone refining (constant melt mass), and pure fractional crystallization (no assimilation). The value of r in any given situation will depend on the balance of heat energy, particularly on the temperature of the country rocks (Table 5.2). The specific heats of the different materials, and the latent heat of crystallization of the minerals on the liquidus, are also relevant. This is because the energy required to heat, melt, and dissolve the country rocks must be derived from the magma. In general, the colder the country rocks, the larger will be r.

A family of differential equations can now be set up to describe the progressive changes of the magma. It is useful to describe these changes from the point of view of the melt, and particularly in terms of a hierarchy of variables that represent the mass of magma (M_m), the mass of the element of interest in the magma ($M_m C_m$) and the mass of a trace isotope of the element of interest in the magma ($M_m C_m R_m$). Following Taylor (1980) and DePaolo (1981), with minor modifications, the equations that express the change of these quantities in terms of the amounts added and the amounts subtracted at any stage become

Table 5.1 Parameters and Definitions for an AFC Model

Assimilate	Magma	Cumulates
Rate added $= A$	Mass $= M_m$	Rate produced $= rA$
Concentration $= C_a$	Concentration $= C_m$	Concentration $= DC_m$
Isotope ratio $= R_a$	Isotope ratio $= R_m$	Isotope ratio $= \alpha R_m$

Table 5.2 Geologic Parameters for AFC Models

Process	Temperatue of Country Rock	r	D_{Sr}	λ
Fractional crystallization	Cold	∞	> 1	∞
High-level AFC (epizonal)	Cold	> 3.5	> 1	~ 5
Mid-level AFC (mesozonal)	Hot	$1{-}3.25$	> 1	≥ 1
Deep crust AFC (catazonal)	Very hot (melting zone)	$0{-}1$	> 1	~ 1
Upper mantle AFC	Very hot (Plag. unstable)	$0{-}1$	< 1	~ 1

$$\text{Mass of magma:} \quad dM_m/dt = A - Ar \tag{5.6a}$$

$$\text{Mass of element:} \quad d(M_m C_m)dt = AC_a - ArDC_m \tag{5.6b}$$

$$\text{Mass of isotope:} \quad d(M_m C_m R_m)/dt = AC_a R_a - ArDC_m \alpha R_m \tag{5.6c}$$

The various terms in these equations may be understood by inspection of Table 5.1, and particularly by taking successive downward products within each vertical column. Several useful solutions to these equations are provided in the following; others are provided by DePaolo (1981), Taylor and Sheppard, (1986), Fleck and Criss (1985), and Albarede (1995).

5.2.3.3 AFC Equation for Oxygen Isotopes

Of most interest here is the AFC equation for oxygen. For oxygen in common silicates, the concentrations C_m, C_a, and C_{xl} are essentially identical, so that D is unity. The solution to equation 5.6c for this case is (R. E. C., unpublished):

$$\left(\frac{R_m - R_m^i}{R_a/b - R_m^i} \right) = 1 - f^{b/(r-1)} \tag{5.7a}$$

where f is the fraction of remaining magma (M_m/M_m^i), and the other parameters are defined in Table 5.1. The parameter b is defined as

$$b = 1 - r + \alpha r \tag{5.7b}$$

For fractionation factors close to unity, the value for b is also close to unity, and to a good approximation is given by the rth power of α; that is, $b \cong \alpha^r$. More important, b defines the final isotope ratio of the magma through the relation $b = R_a/R_m^f$. In terms of δ-values, this relation becomes

$$b = \frac{1000 + \delta_a}{1000 + \delta_m^f} \tag{5.7c}$$

With this definition, equation 5.7a may be directly translated into δ-values:

$$\left(\frac{\delta_m - \delta_m^i}{\delta_m^f - \delta_m^i} \right) = 1 - f^{b/(r-1)} \tag{5.7d}$$

Hypothetical calculations (Figure 5.3) show the strong dependence of AFC systems on r.

Equations 5.7a and d correctly reduce to the condition for simple fractional crystallization (Rayleigh conditions) as r approaches infinity, and to simple mixing conditions for $r = 0$. Calculations for several different cases are shown in Figure 5.3. While equations 5.7a and d represent an improvement over the approximate expressions in the literature, they nevertheless have the drawback that the parameter f cannot be directly measured. A superior approach is to develop solutions to the differential equations 5.6a, b, and c that directly compare observable geochemical variables. Such solutions can either relate the isotope ratios and the concentrations of a given element, or relate the isotope ratios of two different elements, as shown in the following sections.

5.2.3.4 Isotope–Concentration Equation   Fleck and Criss (1985) developed a useful equation that relates variations in the concentration and the isotope ratio of nonfractionating elements. For this case, $\alpha = 1$, and the combined solution to equations 5.6b and c is

$$\left(\frac{R_m - R_m^i}{R_a - R_m^i}\right) = \frac{\dfrac{1}{C_m} - \dfrac{1}{C_m^i}}{\dfrac{\lambda}{C_a} - \dfrac{1}{C_m^i}} \tag{5.8a}$$

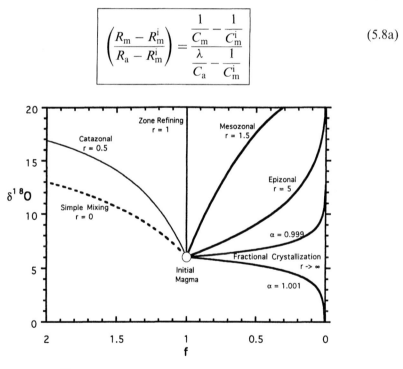

Figure 5.3 Calculated $\delta^{18}O$ values for a melt, with an initial value of $+6.0$, that undergoes AFC interactions with country rock with a $\delta^{18}O$ value of $+20.0$ (equation 5.7a). The AFC trends sweep clockwise with increasing r, generally coincident with a change from catazonal to mesozonal to epizonal emplacement conditions. All curves assumed a crystal–melt fractionation factor of 1.00, except for the fractional crystallization curves specifically labeled.

where the AFC parameter λ is defined as

$$\lambda = 1 + r(D - 1) \tag{5.8b}$$

Fleck and Criss showed that on a plot of the isotope ratio versus the inverse concentration of the element of interest, this equation produces straight lines that originate at the magmatic endmember and that project to the point $(R_a, \lambda/C_a)$.

For Sr isotopes, the left-hand side of equation 5.8a can be expressed to good approximation in terms of $^{87}Sr/^{86}Sr$ ratios. The accuracy of this type of substitution depends on the isotope pair in question, and considerable error will result for certain elements. Nevertheless, for the case at hand, equation 5.8a becomes to very good accuracy (Fleck and Criss, 1985):

$$\frac{[^{87}Sr/^{86}Sr]_m - [^{87}Sr/^{86}Sr]_i}{[^{87}Sr/^{86}Sr]_a - [^{87}Sr/^{86}Sr]_i} \simeq \frac{[1/Sr]_m - [1/Sr]_i}{[\lambda/Sr]_a - [1/Sr]_i} \tag{5.8c}$$

Graphs of the isotope ratio $^{87}Sr/^{86}Sr$ versus the inverse concentration of strontium, that is, $1/[Sr]$, are of great practical importance to geochemists, mostly because binary mixing processes produce linear trends on such plots. As shown by Fleck and Criss (1985), AFC processes also produce linear relationships on such a plot (Figure 5.4; equation 5.8c). As an example, consider the assimilation of ancient, radiogenic country rock by a high-Sr, mantle-derived magma. A family of AFC lines is defined that radiate from the point that represents the magmatic endmember, and whose slopes depend on the endmember compositions and on the value of λ. Only in the case where $\lambda = 1$ do the AFC lines project to the composition of the assimilated country rock (Figure 5.4). This latter situation will occur either for the case where $D = 1$, or for the well-known case of simple mixing ($r = 0$; see equation 5.8b).

Fleck and Criss (1985) also showed that, for the case of simple mixing ($r = 0$), or for any AFC process where $D_{Rb} = D_{Sr}$, a "pseudoisochron" is defined where

$$\frac{[^{87}Sr/^{86}Sr]_m - [^{87}Sr/^{86}Sr]_i}{[^{87}Sr/^{86}Sr]_a - [^{87}Sr/^{86}Sr]_i} = \frac{[Rb/Sr]_m - [Rb/Sr]_i}{[Rb/Sr]_a - [Rb/Sr]_i} \tag{5.8d}$$

On a plot of the $^{87}Sr/^{86}Sr$ versus the Rb/Sr ratio, it is possible for such a suite of rocks to define linear trends. The parameters of this graph are similar to those of the well-known isochron plot from which geochronologic ages are calculated. Equation 5.8d shows that the slopes of such trends can owe their origin to the mixing of distinct endmembers, rather than to elapsed time. Fortunately, in geological systems, the chemical variability of the endmembers is typically rather large, so "pseudoisochrons" produced by mixing are typically poorly defined compared with good isochrons.

5.2.3.5 Isotope–Isotope Equation   Another useful application is to compare the impact of an AFC process on the isotope ratios of two different elements. For two nonfractionating elements (Criss, in Taylor and Sheppard, 1986):

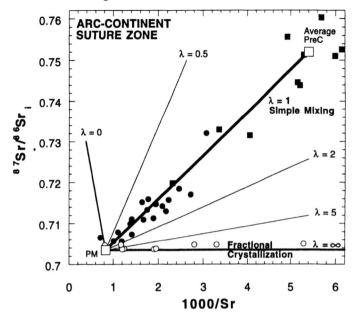

Figure 5.4 Graph of $^{87}Sr/^{86}Sr$ ratios, representing either initial ratios or calculated ratios 80 Ma ago, versus the inverse Sr concentration for rocks near the arc-continent suture zone in western Idaho. Two different suites of granitic rocks are seen: an older suite (open circles) that exhibits a fractional crystallization trend, and a suite (filled circles) that formed during the Cretaceous collisional event that exhibits a steep slope on the diagram. The latter plutons appear to represent mixtures of two fundamentally different source materials (open squares), a primitive magma ("PM") from the arc, and average Precambrian wall-rock ("Average PreC") derived from the continent. Individual samples of Precambrian wallrocks (filled squares), and particularly the average value for these rocks, lie right along the trend line defined by the magmas, indicating that a simple mixing process is involved. For comparison, other AFC trends for different values of λ are shown—these trends sweep clockwise as λ increases. Data from Fleck and Criss (1985).

$$\left\{ \left(\frac{1}{\gamma} - 1 \right) \frac{\lambda C_m^i}{C_a} + 1 \right\}_{\text{element A}}^{1/\lambda} = \left\{ \left(\frac{1}{\gamma} - 1 \right) \frac{\lambda C_m^i}{C_a} + 1 \right\}_{\text{element B}}^{1/\lambda} \qquad (5.9a)$$

where the parameter γ is defined as

$$\gamma = \left(\frac{R_m - R_a}{R_m^i - R_a} \right) \qquad (5.9b)$$

This solution reduces to a simple mixing equation for the case where $r = 0$.

A similar version of equation 5.9 may be developed to compare stable and radiogenic isotopes. This is useful because the isotopic compositions of the end-members are produced by different processes; that is, by the respective processes of fractionation and radioactive decay. For strontium versus oxygen isotopes, the result is (R. E. C., unpublished):

$$\gamma_{oxygen} = \left\{\left(\frac{1}{\gamma_{Sr}} - 1\right)\frac{\lambda C_m^i}{C_a} + 1\right\}^{-b/\lambda} \tag{5.9c}$$

where b and λ are as defined previously, γ_{Sr} is defined in equation 5.9b, and γ_{oxygen} is also given by equation 5.9b, except that for oxygen R_a must be replaced by R_a/b, which is a quantity identical to R_m^f.

5.2.3.6 Example: Adamello Batholith   The best documented example of a magmatic suite that underwent an AFC process is the Adamello batholith in the southern Alps of northern Italy. This composite granitic batholith, the type area for a common granitic rock called adamellite, also includes tonalite and quartz diorite plutons that were all intruded about 35–40 Ma (Cortecci et al., 1979).

An AFC model for this composite batholith (Figure 5.5), similar to that developed by Taylor (1980), shows good agreement with the data of Cortecci et al. (1979). The magmatic endmember is inferred to be an alkali basalt with a $\delta^{18}O$ value of $+5.7$, and a concentration of 900 ppm of Sr that has an $^{87}Sr/^{86}Sr$ ratio of 0.7039. The other endmember, representing pelitic metasedimentary country rocks that were assimilated by the magma, is estimated to have had an average $\delta^{18}O$ value of $+14.0$, and a concentration of 150 ppm of Sr with an $^{87}Sr/^{86}Sr$ ratio of 0.736. Except for the two gabbro samples at the lower end of the line, the data conform reasonably well to the AFC calculations, based on these endmembers and a value for λ of 3.0 (Figure 5.5). The concavity of the strontium versus the oxygen isotopic trend is typical of crustal contamination of primitive magma.

On correlation diagrams of $\delta^{18}O$ versus $^{87}Sr/^{86}Sr$, many igneous rock suites define positive trends. The majority of these appear to radiate from a mantle or MOR basalt endmember with a $\delta^{18}O$ value close to $+5.7$ and a $^{87}Sr/^{86}Sr$ ratio near 0.703 (Taylor and Sheppard, 1986). However, most of these trends are nearly linear, and, except for Adamello, do not display the strong curvature that is predicted by AFC models.

5.3 Igneous Rocks: Subsolidus Processes

5.3.1 Basic Definitions

The variability of whole-rock oxygen and hydrogen isotope ratios in terrestrial igneous rocks is larger than that for magmas. Moreover, the observed patterns of distribution of these isotopes among coexisting minerals are complex and commonly cannot be explained in terms of magmatic fractionation processes. The available data require that subsolidus (secondary) processes were critically important in the generation of isotopic patterns in numerous igneous rock suites. Oxygen isotope data clearly demonstrate the importance and complexity of subsolidus processes in igneous rocks.

Mathematical models of subsolidus processes are primarily based on mass conservation relationships for the various isotopes. These laws are most usefully expressed in terms of closed and open rock systems.

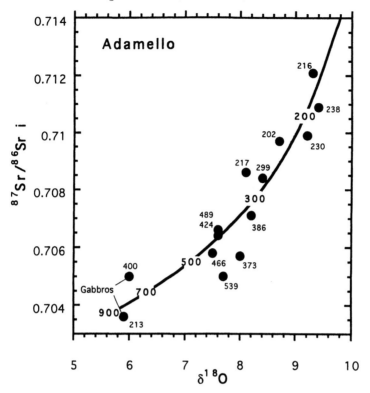

Figure 5.5 Graph of initial $^{87}Sr/^{86}Sr$ ratios versus the $\delta^{18}O$ values of granitic rocks form the Adamello massif, Italy, measured by Cortecci et al. (1979). An AFC model, slightly modified after that of Taylor (1980), is shown by the heavy line, along which calculated Sr concentrations are indicated in larger numbers. The Sr concentrations measured in the individual samples (filled circles) are indicated by the small numbers, and, except for the two gabbroic samples, agree well with the model. Endmembers are given in the text, and the value of 3.0 for the parameter λ was defined by the Sr-isotope and concentration data.

Closed isotopic systems feature one or more coexisting phases in an environment in which heat but not mass can be transported through the system boundaries, whether the latter be real or imaginary. For oxygen and hydrogen isotopes, this condition is generally satisfied if no fugitive phase, such as a fluid, gas, or melt, can enter or exit the system. In the simplest case, only solid phases are present, and no potentially mobile phase is present at all. In the more general case of a "closed" system, a fluid phase can coexist with the solids, but the fluid must forever remain in contact only with those other phases that constitute the system.

Open isotopic systems feature a matrix of coexisting solid phases plus a mobile phase that may infiltrate into, interact with, and pass out of the system. Such systems are therefore characterized by a flux of mass, as well as heat, and consequently by changes in bulk composition over time. In the limiting case of a "buffered" open system, the fluid flux rate is very fast compared with the rate of reaction and isotopic exchange within the system.

The following development will examine closed and open isotopic systems from the perspective afforded by δ–δ diagrams. This approach directly links mathematical models with observable isotopic data, with a minimum of unconstrained system parameters. Figure 5.6 illustrates several different relationships on such plots that will be described in the following sections.

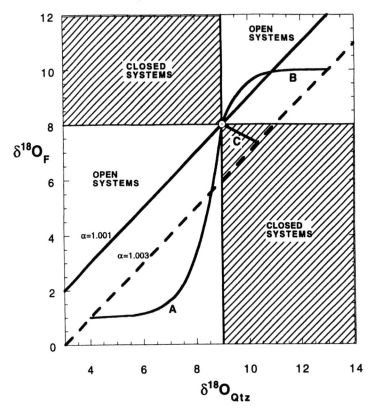

Figure 5.6 Relationships on a δ–δ plot for a bimineralic rock, in this case for a granitic rock composed essentially of quartz and feldspar. Isotherms for a high temperature (solid line, $\alpha = 1.001$) and for a moderate temperature (dashed line, $\alpha = 1.003$) are shown, defining lines of slope α that approximate straight, 45°-sloped lines on the diagram. If a rock, initially at isotopic equilibrium at the higher temperature (open circle), cools to the lower temperature, isotopic exchange can proceed along different trajectories as a new state of equilibrium is approached. Closed system exchange must follow a fixed slope defined by the ratio of mole fractions, $-X_{qtz}/X_F$, as shown by trend "C" for a rock with a 2:1 feldspar-to-quartz ratio. Regardless of the temperature change, the mole fractions, or the exchange mechanisms, all closed system trajectories must be negative, and therefore the bimineralic rock must move into one of the diagonally ruled areas on the diagram. In contrast, open system exchange will typically occur along trajectories with positive slopes, as illustrated for the buffered open system exchange of the rock with an infiltrating, isotopically depleted meteoric fluid (trend "A") or with an isotopically enriched fluid (trend "B"). See text.

5.3.2 Closed Systems

In a closed system, the bulk composition is preserved over time. For example, for such a system constituted of n components, mass conservation requires that the whole-rock $^{18}O/^{16}O$ ratio is given by the well-known and virtually exact relation

$$R_{wr} = \sum_{i=1}^{n} X_i R_i \tag{5.10}$$

where the X_i and R_i, respectively, represent the mole fraction and isotope ratio of each individual phase for the element of interest. As discussed in chapter 2, this simple approximation is virtually exact for oxygen and hydrogen isotopes, as well as for many others, at their natural levels of abundance, so it has numerous applications (Gregory et al., 1989).

If two or more different samples are mixed in a bulk ratio defined by the X_i, then R_{wr} simply becomes the bulk isotopic ratio of the resultant mixture. For a redistribution of isotopes among the different phases that constitute a closed isotopic system, the value of R_{wr} will be invariant even though the individual R_i values may all vary. Important additional constraints on such systems are provided by the principles of isotopic equilibrium, or, in transient systems, of isotopic exchange.

5.3.2.1 Equilibrated Closed Systems   For the case where isotopic equilibrium is attained among all n phases that constitute a closed system, the isotope ratios of all phases are completely constrained. This is because an equilibrium relationship must be satisfied for every pair of phases, giving $n - 1$ independent equations of the form

$$\alpha_{AB} = R_A/R_B \tag{5.11a}$$

In a closed system, the bulk composition is preserved over time. For example, for such a system comprising only two phases, A and B, equation 5.10 becomes

$$R_{wr} = X_A R_A + X_B R_B \tag{5.11b}$$

Because X_B is equal to $1 - X_A$ for a binary system, the relationship defined by equations 5.11a and b is

$$R_B = \frac{R_{wr}}{1 + X_A(\alpha_{AB} - 1)} \tag{5.11c}$$

For the general case of a system composed of n phases, the relationship defined by equation 5.10 and $n - 1$ equations of isotopic equilibrium is

$$R_k = \frac{R_{wr}}{\sum_{i=1}^{n} \alpha_{ik} X_i} \tag{5.12}$$

Of course, the isotope ratios in equations 5.11c and 5.12 may be easily translated into δ-values using the appendix relationships.

5.3.2.2 Slow Cooling or Heating   A fundamental problem involves the prediction of the path that a rock sweeps out during closed system equilibration over a temperature range. In this case, the whole-rock isotope ratio for a rock with n phases is constant and is given by equation 5.10.

The fractionation factor α between any two mineral phases, j and k, usually varies with temperature (T, in kelvins) as

$$\ln \alpha_{jk} = \frac{A_{jk}}{T^2} + B_{jk} \tag{5.13}$$

where A_{jk} and B_{jk} are constants. Given the above, and assuming that isotopic equilibrium is maintained, a closed rock system subject to a temperature variation is constrained to move along trends that have slopes of (Criss and Gregory, 1990):

$$\left(\frac{\partial R_j}{\partial R_k}\right)_{R_{\text{wr}}, X_i} = \frac{\alpha_{jk} \displaystyle\sum_{i=1}^{n} \alpha_{ij} A_{ij} X_i}{\alpha_{kj} \displaystyle\sum_{i=1}^{n} \alpha_{ik} A_{ik} X_i} \tag{5.14}$$

This slope is identical to the slope $\partial \delta_j / \partial \delta_k$ on a δ–δ plot. For systems with three or more phases (i.e., for n > 2), the slope will vary slightly with temperature because of the dependence on the α values. However, in a two-phase system, the slope has a constant value of $-X_k/X_j$, consistent with the relationship obtained by differentiating equation 5.11b. A closed system trend of this type is shown in Figure 5.6 (trend "C").

Failure of rock systems to obey the equilibrium and mass balance constraints is definitive evidence for isotopic disequilibrium and/or open system behavior. Evidence for the latter processes is commonly observed in slowly cooled or infiltrated rocks (e.g., Giletti, 1986; Gregory and Criss, 1986).

5.3.2.3 Kinetic Effects in Binary Closed Systems   Isotopic exchange effects in transient closed systems may be described by combining a material balance relationship with equations that describe isotopic exchange. The simplest and most useful of the latter is the simple kinetic relationship developed in chapter 4. As already discussed, the rate of isotopic exchange is proportional to the deviation of the instantaneous isotope ratio from the equilibrium ratio. Thus, for a binary system composed only of phases A and B, whose respective isotope ratios R_A and R_B would define a quotient equal to α_{AB} at equilibrium, the rate is (Criss et al., 1987)

$$\frac{dR_A}{dt} = k_{AB}(R_A - \alpha_{AB}R_B) \tag{5.15a}$$

For such a binary closed system, it may be shown by differentiating equation 5.11b that

$$X_A \, dR_A = -X_B \, dR_B \tag{5.15b}$$

These two differential equations have the solution

$$\ln\left(\frac{R_A - R_A^{eq}}{R_A^i - R_A^{eq}}\right) = -\left(\frac{\alpha_{AB} X_A + X_B}{X_B}\right) kt \tag{5.15c}$$

where it has been assumed that k is a constant. This expression may be recast in terms of the fractional approach to equilibrium, F:

$$\ln(1 - F) = -\left(\frac{\alpha_{AB} X_A + X_B}{X_B}\right) kt \tag{5.15d}$$

For a binary system, the value of F at any stage must be the same for each phase. Thus, the $1 - F$ term for one phase must be equal to that of the other, giving

$$\left(\frac{R_A - R_A^{eq}}{R_A^i - R_A^{eq}}\right) = \left(\frac{R_B - R_B^{eq}}{R_B^i - R_B^{eq}}\right) \tag{5.16}$$

Note that equation 5.16 does not depend upon any particular mechanism of exchange, but rather is a necessary consequence of material balance (equation 5.10).

5.3.2.4 Corrected Partial Exchange Equation   A useful application of these results may be made to the experimental determination of fractionation factors in systems that have not gone all the way to equilibrium. This "partial exchange" technique has been widely used to determine isotopic fractionation factors because, for several systems, particularly ones at low temperatures, the isotopic exchange rates are too slow to permit complete equilibrium to be achieved on reasonable laboratory timescales. Fractionation factors between various minerals and water are of particular importance, so the following expressions, though general for any two phases A and B, are here written in terms of a mineral phase A and water W.

In the "partial exchange" technique, the equilibrium fractionation factor α^{eq} between the mineral and water may be defined in terms of two other "fractiona-tions." The first, α^i, represents the ratio R_A^i / R_W^i at the beginning of an exchange experiment. The second, α^f, represents the ratio R_A^f / R_W^f at any "final" time, generally the time when the experiment is terminated and the materials are mea-sured. In their original description of the partial exchange technique, Northrop and Clayton (1966) proposed that data from incompletely equilibrated systems could be analyzed on a graph of $\ln \alpha^i$ plotted against the difference $(\ln \alpha^f - \ln \alpha^i)$. For a series of experiments, identical in all respects except for the value of the α^i, such a plot will define a trend whose y-intercept yields the desired value of α^{eq} (Northrop and Clayton, 1966).

Criss et al. (1987) derived the following expression that relates the previously defined quantities in a binary closed system:

$$\boxed{1 - F = \left(\frac{R_W^f}{R_W^i}\right)\left(\frac{\alpha_{A-W}^f - \alpha_{A-W}^{eq}}{\alpha_{A-W}^i - \alpha_{A-W}^{eq}}\right)} \tag{5.17}$$

This equation may be easily recast in terms of the familiar parameters of the partial exchange technique:

$$\alpha^i_{AW} = \frac{1}{G-1}(\alpha^f_{A-W} - \alpha^i_{A-W}) + \alpha^{eq}_{A-W} \qquad (5.18a)$$

where

$$G = \frac{R^i_W}{R^f_W}(1 - F) \qquad (5.18b)$$

Equations 5.18a and b differ from the currently employed partial exchange model in several subtle but important ways. First, Northrop and Clayton (1966) derived their result in terms of a specific kinetic model. In contrast, like equation 5.17, equations 5.18a and b apply to any closed binary system, as they are a direct and necessary mathematical consequence of material balance (equation 5.11b).

Second, equation 5.18b defines the conditions required for this equation to be linear for a series of experiments—that is, G must be a constant. The slope depends not only on F, but also in a critical and complex manner upon the relative mole fractions of the mineral and the water. Moreover, equation 5.18a denotes a linear relationship between the α values, not between the logarithms of the α values. Use of logarithms in the conventional equation introduces an unnecessary approximation that degrades the accuracy of equations 5.18a and b, which are simple, virtually exact expressions for trace isotopes.

Third, and most important, for a set of experiments conducted under conditions of differing α^i but identical G, a graph of α^i, plotted against the difference $(\alpha^f - \alpha^i)$, will yield a linear relationship. Equation 5.18a indicates that this line will have a slope of $(G-1)^{-1}$ and a y-intercept that is identical to the desired value of α^{eq}. Equation 5.18b precisely defines the requirements for G to be the same for the various experimental runs. First, this factor differs from that proposed by Northrop and Clayton (1966), whose analysis is incorrect and is based on extensive, rather than intensive, variables. Second, equation 5.18b indicates that not only must the parameter F be the same, but the quantity R^i_W/R^f_W must also be the same, for all the experimental runs defining the trend. In addition, for many exchange processes, including kinetic exchange (equation 5.15d) and diffusional exchange in a limited volume (Crank, 1975), the quantity F depends in a complicated way on the relative mole fractions of the mineral and water. The best way to ensure the applicability of the partial exchange method is to conduct all the experiments at very high water-to-mineral ratios, so that X_W approaches unity. Only for this case will the slope of the trend reduce to the value of $-1/F$.

Equations 5.18a and b suggest a corrected method to evaluate to partial exchange data. First, for pairs of experiments denoted by subscripts 1 and 2, the $1 - F$ terms of equation 5.17 may be equated, resulting in a quadratic expression that may be readily solved for α^{eq}. Thus:

$$(1 - H)(\alpha^{eq}_{A-W})^2 + [H\alpha^i_1 + H\alpha^f_2 - \alpha^i_2 - \alpha^f_1](\alpha^{eq}_{A-W}) + [\alpha^f_1\alpha^i_2 - H\alpha^f_2\alpha^i_1] = 0 \qquad (5.18c)$$

where

$$H = \left(\frac{R^i_W}{R^f_W}\right)_1 \left(\frac{R^f_W}{R^i_W}\right)_2 \tag{5.18d}$$

Of course, for the simple case where $H = 1$:

$$\alpha^{eq}_{A-W} = \frac{[\alpha^f_2\alpha^i_1 - \alpha^f_1\alpha^i_2]}{[(\alpha^f_2 - \alpha^i_2) - (\alpha^f_1 - \alpha^i_1)]} \tag{5.18e}$$

Alternatively, for a set of data a plot is constructed of α^i versus the difference $(\alpha^f - \alpha^i)$. The y-intercept of the linear regression gives an estimate for α^{eq}, and the inverse of the slope provides an estimate for $-F$ for the set. Then, each individual value for $(\alpha^f - \alpha^i)$ is multiplied by a correction factor:

$$\text{Correction factor} = \frac{F(R^f_W/R^i_W)}{F - 1 + (R^f_W/R^i_W)} \tag{5.18f}$$

where the measured values for the initial and final waters are used for each run. Reconstruction of the plot after this correction will lead to a revised value for the equilibrium fractionation factor and for F, which may be used in a further iteration if necessary.

Perhaps an even better experimental method for the determination of isotopic fractionation factors would be to exchange minute quantities of water, CO_2, or O_2 with a large amount of mineral, so that X_A approaches unity. According to equation 5.15d, this procedure will greatly increase the value of F for a given amount of time, so that a greater approach toward the equilibrium state is realized. Of course, in this method it is the water or the gas, rather than the mineral, that would be analyzed to determine F. Gas–solid exchange experiments, such as those of Muehlenbachs and Kushiro (1974), naturally take advantage of this material balance effect.

5.3.2.5 Kinetic Effects in Multiphase Closed Systems   Transient effects in multiphase closed systems may be treated in much the same manner as above, except that there are more variables and more constraining equations. For an n-phase system, equation 5.10 may be used in conjunction with $n - 1$ independent exchange equations of the form:

$$\frac{dR_i}{dt} = k_{ij}(R_i - \alpha_{ij}R_j) \tag{5.19}$$

As an example, Criss et al. (1987) obtained a solution for a three-phase system, defined in that case as two different minerals (A and B) plus aqueous fluid (W) that all exchange in a closed system. The material balance relationship, obtained by differentiating equation 5.10, is

$$X_A\,dR_A + X_B\,dR_B + X_W\,dR_W = 0 \tag{5.20a}$$

The kinetic equations are

$$\frac{dR_A}{dt} = k_{AW}(R_A - \alpha_{A-W}R_W)$$ (5.20b)

and

$$\frac{dR_B}{dt} = k_{BW}(R_B - \alpha_{B-W}R_W)$$ (5.20c)

Criss et al. (1987) found an analytical solution that simultaneously satisfies these three differential equations, 5.20a, b, and c:

$$\begin{bmatrix} R_A \\ R_B \\ R_W \end{bmatrix} = \begin{bmatrix} \dfrac{k_{AW}\alpha_{A-W}}{k_{AW}-\lambda_1} & \dfrac{k_{AW}\alpha_{A-W}}{k_{AW}-\lambda_2} & R_1^{eq} \\ \dfrac{k_{BW}\alpha_{B-W}}{k_{BW}-\lambda_1} & \dfrac{k_{BW}\alpha_{BW}}{k_{BW}-\lambda_2} & R_2^{eq} \\ 1 & 1 & R_W^{eq} \end{bmatrix} \times \begin{bmatrix} e^{-\lambda_1 t} & 0 & 0 \\ 0 & e^{-\lambda_2 t} & 0 \\ 0 & 0 & 1 \end{bmatrix} \times \begin{bmatrix} C_1 \\ C_2 \\ 1 \end{bmatrix}$$

(5.21a)

where the eigenvalues (λ_1, λ_2) are given by the roots of the equation

$$X_W\lambda^2 - [k_{AW}(\alpha_{A-W}X_A + X_W) + k_{BW}(\alpha_{B-W}X_B + X_W)]\lambda$$
$$+ k_{AW}k_{BW}(\alpha_{A-W}X_A + \alpha_{B-W}X_B + X_W) = 0$$ (5.21b)

The constants C_1 and C_2 may be evaluated from the initial conditions.

For each additional phase C, an additional term $X_c dR_c$ would need to be incorporated into equation 5.20a, and another kinetic equation of the form of equations 5.20b and c would need to be added to the set. The differential equations for this case and for more complex cases that involve additional phases may be solved by extension of the matrix, or perhaps more readily by iterative calculations on a computer.

5.3.3 Open Systems

Isotopic effects in open systems may also be understood in terms of material balance and either equilibrium or kinetic equations. For such systems, the most realistic scenario is that of a fluid phase that enters into, interacts with, and passes out of a matrix of phases, usually solids, that are confined within the system. This infiltrating, fugitive phase may be an aqueous fluid, or a gas, or magma, and so on. The interactions of this fugitive phase with the other phases in the system will cause complex variations in composition of every phase over time. Unlike a closed system, the whole-rock or "bulk" composition of an open system will also vary.

5.3.3.1 Equilibrated Open Systems   With the passage of time, in an open system, the infiltrating phase will ultimately dominate the material balance relationships. At this final stage, the infiltrating phase will not change in composition as it passes through the system, and equilibrium will be established between that fluid and all other phases. In this case, the isotopic ratio of the fluid phase is identical to that of the entering fluid, R_w^{in}. The isotope ratio R_A of all the other phases in the system are then simply defined by the family of equations of the form

$$\alpha_{AW}^{\text{eq}} = R_A / R_W^{\text{in}} \tag{5.22}$$

5.3.3.2 Buffered Open Systems   In another useful case, the "buffered" open system, the infiltrating phase passes through the system so rapidly that it does not sensibly change in composition during transit. Over time, however, the solid matrix phases attempt to attain a composition in equilibrium with the dominant fluid. The rates of exchange of the minerals are all defined by equations of the form of equation 5.19, one for each separate phase, but with the value for R_W being replaced by the constant value R_W^{in}. Each exchange equation of this type behaves independently of the others in such a buffered system. Thus, the solutions are straightforward, and, for each phase, have the form

$$\ln \left(\frac{R_A - \alpha_{AW} R_W^{\text{in}}}{R_A^{\text{i}} - \alpha_{AW} R_W^{\text{in}}} \right) = -kt \tag{5.23}$$

Note that this result is the same as that for a closed system in which the water reservoir is effectively infinite—that is, $X_w \to 1$ (cf. equation 5.15c).

It is sometimes very useful to eliminate time as an explicit variable. This may be accomplished by comparing the isotope ratio of two coexisting solid phases in the system. For example,

$$\left(\frac{R_A - R_A^{\text{eq}}}{R_A^{\text{i}} - R_A^{\text{eq}}} \right) = \left(\frac{R_B - R_B^{\text{eq}}}{R_B^{\text{i}} - R_B^{\text{eq}}} \right)^{k_{AW}/k_{BW}} \tag{5.24a}$$

which is identical to

$$(1 - F_A) = (1 - F_B)^{k_{AW}/k_{BW}} \tag{5.24b}$$

Such expressions are important in the study of ancient processes because they directly compare the isotopic ratios of two materials that, in many cases, can be directly measured (e.g., Figure 5.6, trends "A" and "B").

5.3.3.3 Transient Open Systems   In the most general case, the isotope ratios of all phases, including the fugitive phase, will all change as a complex function of time as an open system attempts to achieve equilibrium. Fundamentally, the approach for a mathematical description is the same; that is, the system can be described in terms of material balance equations and a set of $n - 1$ independent equations that describe the transient effects for each phase. A modified material balance equation, plus a set of kinetic equations of the form of equations 5.20b and c, was analytically solved by Criss et al. (1987). Alternatively, such differential equations may be readily solved by numerical iteration, even for systems that contain a large number of phases.

5.3.4 The δ–δ Plot

The most straightforward way to interpret isotopic data from complex natural systems is to graph the results on a δ–δ plot. Such plots simply depict the measured δ-values of one phase versus those of a coexisting phase. For an n-phase system, such plots are simple projections of an n-dimensional construct called delta space, whose mathematical properties have been discussed by Gregory and Criss (1986).

On a δ–δ plot, many complex processes produce linear trends (Figure 5.6). First, for an equilibrated system, various isotherms will approximately define a set of parallel lines with the slopes α_{AB}^{eq} appropriate for the different temperatures (equation 1.25). Of course, these isotherms will have approximately unit slopes and intercepts of Δ_{A-B}. Several other graphs, such as δ–Δ plots or Δ–Δ plots, are also commonly used to interpret isotopic data, but these convolute the basic data and are subject to undesirable induced correlations. Besides, as shown by Gregory and Criss (1986), any linear trend on such plots will mathematically transform into a linear trend on a δ–δ plot.

A particularly useful feature of δ–δ plots is the simple manner in which closed and open systems can be distinguished (Figure 5.6). This is most easily illustrated for a system composed of two solid phases. Many igneous rocks approximate bimineralic assemblages, as they are $>90\%$ composed of two minerals; for example, granites (quartz and feldspar), gabbroes (plagioclase and pyroxene), and peridotites (olivine and pyroxene). Gregory et al. (1989) provide numerous natural examples.

For a binary closed system, isotopic exchange must produce a negative slope on a δ–δ plot. This is because any gain of a heavy isotope by one phase must be balanced by the loss of the heavy isotope in the other. This simple situation is evident in the material balance equation (equation 5.15b), which further shows that the slope of a closed system exchange trend must be equal to $-X_A/X_B$. Moreover, since the mole fractions are necessarily positive real numbers, this latter quantity is invariably negative (Figure 5.6).

In the case where three or more phases undergo exchange over a temperature range in a closed system, the slope on a δ–δ plot that depicts any pair may be calculated from equation 5.14, provided that equilibrium is maintained. These slopes are generally, but not always, negative. For the special case where the δ–δ plot depicts the most ^{18}O-rich mineral versus the most ^{18}O-poor mineral, the slope will be invariably negative, regardless of the complexity of the assemblage or the magnitudes of the various mole fractions (Gregory and Criss, 1986).

Open systems exhibit a contrasting behavior on δ–δ plots (Figure 5.6). In open systems, the exchange process does not merely involve the transfer of isotopes among the solid phases in the system. Instead, the passage of an oxygen-bearing fluid induces mineral–fluid exchange, and also causes a progressive change in the whole-rock isotopic ratio. The isotopic response of such systems is complex and involves many factors, but most important are the initial isotope ratios of the various minerals, the ratios that the various minerals would have were they at equilibrium with the infiltrating fluid, and also the mineral–fluid exchange rates and the amount of fluid that has transited the system up to any given time. The initial and equilibrium isotopic compositions define two points on a δ–δ diagram that, respectively, constitute the ends of a generally curved line segment that represents the trajectory of isotopic exchange. The position along this trajectory is a function of time and depends on the various exchange rates and other factors.

In typical open systems, the δ-values of the minerals will either all progressively decrease, or, less commonly, all increase, in response to progressive fluid infiltration. This is because the δ-value of the infiltrating fluid will commonly have an "exotic" value, far from equilibrium with the particular rock. For this reason, open systems typically exhibit positive slopes on δ–δ diagrams. Such positive-sloped trends are, in fact, the diagnostic signature of open system behavior, because for major phases they are practically impossible to generate by closed system exchange.

5.4 Hydrothermal Systems

5.4.1 Water–Rock Interactions

Because of its extensive hydrosphere and dynamic interior, fluid–rock interactions are commonplace on Earth. Oxygen and hydrogen isotope data prove that such interactions have affected most rocks of Earth's crust and upper mantle.

The $\delta^{18}O$ and δD values of rocks that interact with significant amounts of hydrothermal fluids can be modified extensively. Typical alteration temperatures range up to the critical point of water, near 374°C, but higher in saline solutions. Much higher temperatures can also occur; for example, in fumaroles. Isotopic exchange between the minerals and the fluids is promoted in all these systems, and a variety of alteration assemblages can be concurrently formed as the systems adjust toward thermodynamic equilibrium under these conditions. In most cases, the interactions involve surface-derived fluids, most commonly meteoric water in continental systems, and seawater in submarine systems. However, magmatic, formation, metamorphic, and other fluids participate in some hydrothermal environments.

Stable isotopes provide unique insight into the characteristics of hydrothermal systems. The systematics offered by coupled oxygen and hydrogen isotope data are particularly powerful, because aqueous fluids are so rich in these elements that they can profoundly modify the O and H isotopes in rocks. For example, graphs of δD versus $\delta^{18}O$ values for altered minerals commonly define characteristic "inverted L" patterns (Taylor, 1977). In addition, on δ–δ plots, the $\delta^{18}O$ values of coexisting minerals from hydrothermally altered rocks define linear, positive-sloped trends that are clearly diagnostic of isotopic disequilibrium and open system (infiltration) histories (Figure 5.7; Gregory et al., 1989). Detailed discussions and examples of these alteration trends are given in Taylor (1977), Criss and Taylor (1983, 1986), and Gregory et al. (1989).

Great insight into the process of hydrothermal alteration may be gained by examination of the process of isotopic exchange, utilizing the same principles discussed earlier. Changes in the ^{18}O contents of rocks during alteration are necessarily balanced against complimentary changes in the ^{18}O contents of the

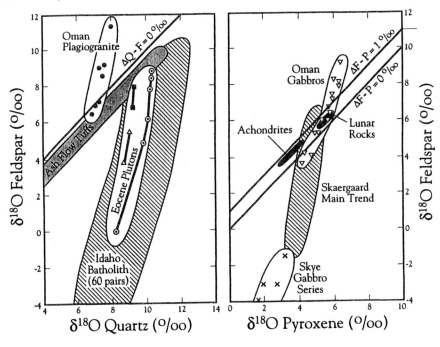

Figure 5.7 A δ-δ plot for quartz–feldspar and for pyroxene–feldspar (plagioclase) pairs from various igneous rock suites. The trends for many granitic and gabbroic rocks define steep positive slopes that clearly indicate exchange under open system hydrothermal conditions. The Idaho batholith, Skaergaard intrusion, and the Skye complex underwent exchange with infiltrating fluids derived from low-^{18}O meteoric waters, while the Oman suites underwent exchange with heated seawater. In contrast, many young ash-flow tuffs, as well as lunar basalts and achondrites, lie along high-temperature isotherms acquired during their magmatic crystallization and preserved by rapid cooling. The low-^{18}O values of some of the ash-flow tuffs probably reflect assimilation of hydrothermally altered roof materials during the caldera collapse that led to their effusion. After Criss (1995), and modified after Gregory et al. (1989).

fluids, as evidenced by the "^{18}O-shift" of geothermal waters, discussed in chapter 3. The most straightforward way to visualize these effects are in terms of H. P. Taylor's water/rock ratios. These ratios are, in effect, the dimensionless numbers that indicate the fundamental character of these exchange systems.

5.4.1.1 Water/Rock Ratios The response of a rock to isotopic exchange with fluid depends on the temperature (e.g., on the rock–fluid fractionation factor), on the exchange rate and event duration, and on the relative proportions and initial isotope ratios R_r^i and R_w^i of the rock and water. In the simplest case, where isotopic equilibrium is attained in a closed system, the conservation relation is (Taylor, 1977)

$$\boxed{\left(\frac{\mathscr{W}}{\mathscr{R}}\right)_{\text{closed}} = -\left(\frac{R_r^f - R_r^i}{R_w^f - R_w^i}\right)} \qquad (5.25)$$

where R_w^f and R_r^f, respectively, represent the "final" equilibrium isotope ratios of the rock and water. Here, the "water/rock ratio," $\mathscr{W}/\mathscr{R}$, defined as the relative molar amounts of the element in question (oxygen or hydrogen), is equal to the ratio of the isotopic shifts that would be observed in the rock and water, analogous to the "lever rule."

Determination of a $\mathscr{W}/\mathscr{R}$ ratio from measurements on an ancient system requires several assumptions. First, while rock specimens may be collected and analyzed, the water in such systems is generally long gone. However, hydrogen isotope data on the altered rocks can commonly be used to estimate the value for R_w^i for oxygen, as discussed by Taylor (1977). The value for R_w^f should, by model assumption, be equal to the quotient R_r^f/α^{eq}, provided that the temperature of exchange can be estimated. Of course, fractionation factors are not determined for "rocks," but, for many igneous rocks, the average "rock–water" fractionation factor will be similar to that for feldspar–water at the given temperature. These matters and additional exchange models are discussed by Taylor (1977), Criss and Taylor (1986), and Gregory et al. (1989).

Few environments of water–rock exchange are closed, so it is useful to develop the $\mathscr{W}/\mathscr{R}$ ratio for an open system. This may be accomplished by assuming that very small packets of water successively enter and completely equilibrate with the rock matrix. Each tiny packet must undergo exchange as governed by equation 5.25, but then it is removed from the system and replaced by another packet. Such a process may be mathematically described by cross-multiplying equation 5.25 and converting it into a differential equation:

$$\left(\frac{R_r}{\alpha_{rw}^{eq}} - R_w^i\right) d\mathscr{W} = -\mathscr{R}\, dR_r \qquad (5.26a)$$

Because the isotope ratio of the incoming packets, R_w^i, is a constant, this result may be easily integrated to yield the open system water/rock ratio:

$$\left(\frac{\mathscr{W}}{\mathscr{R}}\right)_{\text{open}} = \alpha \ln\left[\frac{R_r^i - \alpha R_w^i}{R_r^f - \alpha R_w^i}\right] \qquad (5.26b)$$

The open and closed system water/rock ratios are related by the expression

$$\left(\frac{\mathscr{W}}{\mathscr{R}}\right)_{\text{open}} = \alpha \ln\left[1 + \frac{1}{\alpha}\left(\frac{\mathscr{W}}{\mathscr{R}}\right)_{\text{closed}}\right] \qquad (5.26c)$$

The latter two equations differ from those of Taylor (1977) only in incorporating the factor α_{r-w}^{eq}. Because the rock–water fractionation factor is normally close to unity, this detail makes little practical difference.

A useful comparison may be made of Taylor's open system $\mathscr{W}/\mathscr{R}$ ratio and the buffered open system relationship (equations 5.24a and b). Undered buffered open system conditions, the value $R_w^i \alpha_{rw}^{eq}$ is identical to the ultimate equilibrium value for the rock, after infinite time. Thus, equation 5.26b becomes

$$\left(\frac{\mathscr{W}}{\mathscr{R}}\right)_{\text{open}} = -(\alpha_{\text{rw}}^{\text{eq}}) \ln\left[1 - F\right]$$ (5.26d)

This result implies that

$$\left(\frac{\mathscr{W}}{\mathscr{R}}\right)_{\text{open}} = (\alpha_{\text{rw}}^{\text{eq}})k_{\text{rw}}t$$ (5.26e)

so that the open system $\mathscr{W}/\mathscr{R}$ ratio is directly proportional to time, as required for a steady flow-thru system.

5.4.1.2 Material Fluxes and the $\mathscr{W}/\mathscr{R}$ Ratio

The water/rock ratio is a dimensionless quantity. As such, it is different than two physical quantities with which it is sometimes confused, notably the fluid flux, with units of g/cm^2-s, and the integrated flux, with units of g/cm^2. As pointed out by Criss and Taylor (1986), the seemingly huge quantities of fluid (e.g., 100–5000 kg/cm^2) calculated by Norton and Taylor (1979) to have passed through certain intrusions are completely consistent with moderate $\mathscr{W}/\mathscr{R}$ ratios close to unity. This is because each segment along a fluid pathline must encounter the same fluid that passed through the previous segment, and so on, all along the entire flowpath that may be several kilometers long.

A simple relationship between fluid flux and the $\mathscr{W}/\mathscr{R}$ ratio may be developed as follows. During the lifetime τ of a hydrothermal system, the fluid penetrates through cracks for an effective distance λ, where λ/τ is the actual microscopic velocity of the fluid. Given a fluid flux F in g_{H_2O}/cm^2-s, the total mass of fluid that passes through a rock element of cross section a^2 is

$$\mathscr{W} = Fa^2\tau$$ (5.27a)

The total mass $\mathscr{R}$ of rock in the element affected by the flow is equal to the total volume times the rock density ρ:

$$\mathscr{R} = \rho a^3$$ (5.27b)

Thus, the integrated $\mathscr{W}/\mathscr{R}$ ratio in mass units in the element is defined by the quotient of these equations:

$$(\mathscr{W}/\mathscr{R})_{\text{mass}} = F\tau/(\rho a)$$ (5.27c)

Note that the numerator on the right-hand side is basically the same as the integrated flux. While this relationship is only semiquantitative, it serves to illustrate the fundamental differences between the fluid flux F, the integrated flux $F\tau$, and the $\mathscr{W}/\mathscr{R}$ ratio.

5.4.1.3 Response of Rock Systems

In conformity with the above relationships, the $\delta^{18}O$ and δD values of rocks that interact with significant amounts of hydrothermal fluids can be changed markedly. For example, where low-^{18}O, low-D meteoric waters interact with rocks at typical hydrothermal temperatures (150–350°C), the rocks can become as low as -10 and -200, respectively (see later). The reduction of the rocks in ^{18}O is the material balance counterpart of the "^{18}O-shift" in geothermal waters, discussed in chapter 3, and required by equation 5.25.

However, in such geothermal fluids, a significant "D-shift" is generally absent even though the "^{18}O-shift" may be large, and this important matter requires clarification.

It is very important to note that, for a given interaction, the $\mathcal{W}/\mathcal{R}$ ratio for hydrogen is much larger than the $\mathcal{W}/\mathcal{R}$ ratio for oxygen. This occurs because typical rocks have very high molar contents of oxygen but quite low contents of hydrogen, while water has high contents of both. For this reason, only a small amount of water is required to produce, by exchange, a significant change in the δD value of a rock. This relationship explains the general lack of a significant "D-shift" in many geothermal fluids, as well as the "L-shaped" trends obtained when the δD values are plotted against those of $\delta^{18}O$ in suites of altered rocks (Figure 5.8; Taylor, 1977; Criss and Taylor, 1986).

At very low temperatures ($< 100°C$), interaction with meteoric waters will again reduce the δD values, but, in this case, the $\delta^{18}O$ value of rocks may actually increase because the rock–fluid fractionation factor for ^{18}O is sufficiently large at low T. Such effects are commonly observed in the low-temperature fringes of hydrothermal systems. Even more pronounced effects, due to glass hydration, may be observed where glassy volcanic rocks exchange with meteoric waters at low temperatures. In such cases, the whole-rock $\delta^{18}O$ values can become larger by several per mil than the values for any unaltered phenocrysts, including quartz, that occur in the rocks (e.g., Stuckless and O'Neil, 1973).

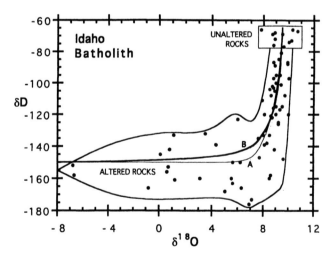

Figure 5.8 Graph of δD values of biotite versus the $\delta^{18}O$ values of coexisting feldspar for samples of the Idaho batholith. The "L-shaped" trend of the data is a product of a profound hydrothermal alteration event during the Eocene time. Exchange with hot infiltrating fluids derived from low-^{18}O, low-D meteoric waters produced large changes from the original isotopic values of the rocks (box), with small amounts of fluid causing large changes in the δD values yet producing little change in the $\delta^{18}O$ values. This effect may be viewed either as the product of the large difference in the water/rock ratios that are calculated on a hydrogen versus oxygen basis (curve A), or a difference in H and O isotopic exchange rates in a buffered open system (curve B), or, alternatively, as a combination of kinetic and material balance effects. Modified after Criss and Taylor (1983).

5.4.2 "Closed" Hydrothermal Systems

Few examples of "closed" hydrothermal systems have been documented. One of the best geological environments for the occurrence of such systems is in the cupolas of certain granitic plutons, which are also zones where valuable ores of Cu, Mo, or other metals can be formed. Greisen deposits, noted for their intense alteration style and their association with Sn-W mineralization, sometimes also form in these environments.

The Silsilah tin deposit in west-central Saudi Arabia is such a deposit. Hot ($\sim$360°C), fluorine-rich aqueous fluids collected beneath aplitic carapaces that developed on top of a crystallizing, highly differentiated granite pluton, producing an impermeable barrier to rising volatiles (Kamilli and Criss, 1996). Intense fluid–rock interaction produced the quartz–topaz–fluorite greisen assemblage immediately beneath these caps. All pre-existing feldspar and mica were destroyed in this zone, and cassiterite (SnO_2), wolframite (($FeMn)WO_4$), and base metal sulfides were formed. Progressing vertically downward from this zone, mica and then feldspars occur along with these minerals, and finally fresh granite is encountered, all over a vertical scale of a few tens of meters.

The $\delta^{18}O$ value of the greisen-stage fluid, about +5.5, may be determined from (1) the measured $\delta^{18}O$ values of the hydrothermal quartz, (2) the temperature of 360°C determined from fluid inclusion measurements, and (3) the quartz–water fractionation factor of Clayton et al. (1972). This fluid appears to be largely derived from the cooling, magmatic hydothermal system, but it also contains a component of externally derived formation or meteoric waters (Kamilli and Criss, 1996). The relative amount of this externally derived fluid progressively increased over the lifetime of this complex hydrothermal system, of which the greisen stage is only a part.

Key insight into the nature of the greisen system may be derived from relationships on a δ–δ plot. The quartz–feldspar samples define a steep, positive-sloped disequilibrium trend that indicates progressive alteration of the granite with the hot, high-^{18}O fluid (Figure 5.9). This trend is mirrored by a negative-sloped trend for the quartz–mica pair. Note that the greisenized samples define the extreme, diverging ends of both trends.

The complementary changes in the ^{18}O contents of the minerals in Figure 5.9 provide compelling evidence that the greisen formed in an essentially "closed" hydrothermal system. A model prediction of the alteration path may be made with equations 5.21a and b. A rock containing equal parts of quartz, feldspar, and mica, with initial $\delta^{18}O$ values of +10.5, 9.5, and 9.0 respectively, is imagined to be placed in a closed system with 10 mol. % fluid with a value of +8.0. These initial $\delta^{18}O$ values typify those measured by Kamilli and Criss (1996) in fresh granites at Silsilah, and, in the case of the fluid, represent the calculated value of the earliest stage fluid in the area. Isotopic exchange then occurs at 350°C, where the quartz–water, feldspar–water, and muscovite–water fractionations are 1.0058, 1.0041, and 1.0027, respectively, producing the trends shown in Figure 5.9. While this model incorporates several simplifying assumptions, note that (1) the fluid quickly evolves to a composition of +5.9, similar to that estimated for the greisen stage; (2) the anomalously small, disequilibrium quartz–feldspar fractionations are

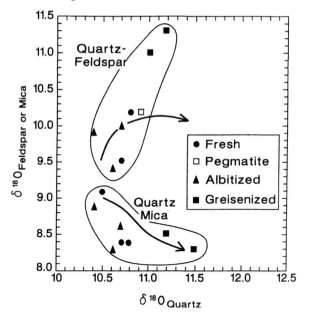

Figure 5.9 A δ–δ plot for quartz–feldspar or quartz–mica pairs for fresh and altered granitic rocks from the Silsilah ring complex, Saudi Arabia. The diverging trends are very
unusual, but compatible with fluid–rock exchange in a closed hydrothermal system, as
illustrated by the calculated trends (arrows). In some of the greisenized samples, the
$\delta^{18}O$ values of feldspar are higher than those of coexisting quartz, representing an isotopic
reversal. From Kamilli and Criss (1996).

reproduced; and (3) the complementary quartz–mica trend is simulated (Figure
5.9). Kamilli and Criss provide additional discussion of this model, along with a
detailed description of the Silsilah deposit.

5.4.3 Oceanic Crust

Significant changes in the $\delta^{18}O$ values of rocks also occur during interactions with
heated seawater. These interactions are particularly pronounced along mid-ocean
ridge (MOR) spreading centers, which are linear belts of magmatic intrusion that
generate the rocks that constitute the ocean floor. An estimated 12 km^3 of basalts
form at MOR spreading centers every year, ultimately to be subducted at a
trench, but this is not all. Beneath the basalts is a thick sequence of layers with
different geophysical properties. These layers are inferred to represent the sheeted
igneous dikes, then gabbroes, and finally the peridotites below the Moho, that
together constitute the thick package known as the ophiolite sequence (Figure
5.10).

The submarine MOR intrusive centers are perfect sites for the generation of
giant hydrothermal systems. Fractures are understandably common in the rocks
near these tectonically active centers, and these are necessarily filled with seawater. This "pore" water will be induced to convectively circulate wherever

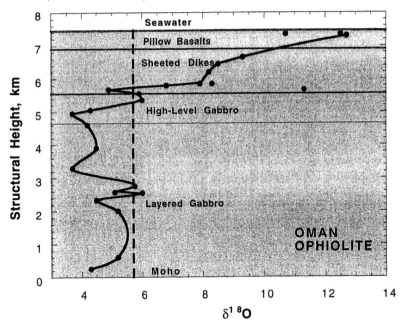

Figure 5.10 Vertical profile of whole-rock $\delta^{18}O$ values for a composite section through the Oman ophiolite, after Gregory and Taylor (1981). The structural height refers to the level above the Moho, below which mantle rocks, such as peridotites, occur, and above which occurs a sequence of rocks inferred to represent an ancient section of seafloor. Hydrothermal alteration at low and high temperatures has respectively produced enrichments and depletions in the $\delta^{18}O$ values of the rocks relative to pristine mantle values, represented by the dashed vertical line at +5.7. Gregory and Taylor argue that the weighted average of such isotopic changes is zero, so that the isotopic composition of the ocean is buffered by seafloor alteration processes over geologic time.

heated by the nearby magmas, so all the ingredients for a hydrothermal system are present. In fact, these systems are so extensive, and so vigorous, that it has been estimated that the entire ocean will pass through a MOR hydrothermal system every $\sim$30 Ma. Moreover, Muehlenbachs and Clayton (1976) generated the important concept that the $\delta^{18}O$ value of seawater is buffered near zero per mil by the extensive water–rock interactions at these centers.

Direct examination of the $\delta^{18}O$ relationships of rocks at great depths below the ocean floor is generally not possible, because drilling is difficult and penetrates only the uppermost zones. In lieu of this approach, rock sequences that are thought to be remnants of oceanic crust that have been thrust upon the continents can be examined. These sequences, known as "ophiolites," have the same vertical sequence of rocks that are thought to constitute the oceanic crust.

The Samail ophiolite, Oman, is probably the largest, best exposed, best preserved ophiolite on Earth. This 700-km-long segment, inferred to have formed during the Cretaceous period in the ancient Tethys seaway, was thrust upon the Arabian peninsula about $\sim$85 Ma ago, according to geochronologic data on biotite from the amphibolite rocks that constitute the basal thrust. Gregory and Taylor (1981) found a large range in the whole rock $\delta^{18}O$ values of this ophiolite, from +3.7 to +12.7, an effect that is attributable to hydrothermal interaction with seawater over a range of temperatures. However, Gregory and Taylor (1981) argue that the weighted average $\delta^{18}O$ value for all these rocks is close to the value of +5.7 inferred for unaltered mantle (Figure 5.10).

The data of Gregory and Taylor provide compelling evidence in support of Muehlenbachs and Clayton's hypothesis. Owing to the variation of the fractionation factor with temperature, interactions between seawater and oceanic rocks below 350°C will increase the $\delta^{18}O$ value of the rock, but at higher temperatures the interactions will cause a reduction in the $\delta^{18}O$ value. The overall $\delta^{18}O$ value of seawater and that of the oceanic crust (taken as a whole), are effectively unchanged by this alteration process when it operates over the entire temperature range. In other words, at the average temperature of interaction of about 350°C, there is a ~6‰ fractionation between the rocks and the water, which is essentially the same as the difference between typical mantle rocks and average ocean water. While many scientists have argued for a change in the oxygen isotope composition of seawater over geologic time, studies of ancient rocks affected by hydrothermal systems suggest otherwise. In particular, the ^{18}O contents of rocks that host massive sulfide deposits, which are clearly the ancient analogues of the spectacular "black smokers," indicate that the oceans have remained nearly constant in ^{18}O for billions of years.

5.4.4 Meteoric-Hydrothermal Systems

Subaerial hydrothermal systems provide the most striking examples of isotopic alteration of igneous rocks. The fluids in these systems are typically dominated by meteoric waters that, in geographically appropriate regions, are highly depleted in D and ^{18}O. Interactions between rocks and such fluids at typical hydrothermal temperatures (150–350°C) commonly produce low $\delta^{18}O$ and δD values that, in extreme cases, become as low as −10 and −200, respectively (Criss and Taylor, 1986). The alteration also produces positive-sloped trends on δ–δ plots that are clearly diagnostic of isotopic disequilibrium and open system (infiltration) histories (Gregory et al., 1989). Moreover, as shown in the following examples, isotopic maps and sections may be made of these systems, and these provide unique insight into the nature of convective fluid circulation on Earth.

5.4.4.1 Stratovolcanic Centers and "Bullseye" Anomalies   One of the most interesting types of isotopic anomalies that have been discovered in Earth's crust are the large, roughly circular, "bullseye" zones of ^{18}O depletion. These anomalies typically occur in andesitic lavas and feature a set of concentric, inwardly decreasing $\delta^{18}O$ contours that encompass areas of 25–150 km^2. Important examples have been documented in many areas, including the Bohemia mining district in the western Cascades of Oregon (Taylor, 1971), the Comstock lode in Nevada, the Yankee Fork district in Idaho, and many other areas. While these examples all occur in Tertiary lavas in the western USA, a "bullseye" anomaly at Pilot Mountain, North Carolina, formed during the lower Paleozoic and survived regional greenschist metamorphism. "Bullseye" isotopic anomalies probably represent the roots of andesitic stratovolcanoes, but direct evidence for such relationships is difficult to obtain because deep erosion normally destroys the familiar conical volcanic edifice before exposing any underlying low-^{18}O zones. However, in the Lassen area of California, Rose et al. (1994) conclusively showed that low-^{18}O rocks lie directly beneath the former site of a Quaternary stratocone whose central zone had been deeply entrenched by a glacial valley.

The Comstock lode region of western Nevada is the best documented area of the "bullseye" type. This famous mining district produced more than 8 million ounces of gold and nearly 200 million ounces of silver from a group of small, high-grade epithermal "bonanza" ore bodies that were deposited in a 10-km-long quartz vein that developed along the Comstock fault. The district is mainly underlain by a thick sequence of Miocene andesites that have undergone pervasive "propylitic" alteration. This alteration is characterized by the development of chlorite, epidote, and other secondary minerals that are now recognized to be the product of meteoric-hydrothermal activity.

By systematic collection and ^{18}O-analysis of rocks, Criss and Champion (1991) made a δ^{18}O contour map that provides much insight into the ancient hydrothermal activity in the Comstock lode district (Figure 5.11). First, the map shows that the field of alteration is very large, comprising a circular, 75-km^2 zone of ^{18}O depletion centered on a 2-km^2 granodiorite stock at Mt. Davidson. The central position of this stock, and the large disparity between the area of the stock and its alteration field, suggest that this small intrusion probably represents the central vent of the main volcanic center.

The geometric regularity of the isotopic anomaly in Figure 5.11 is clearly disrupted by the Comstock fault, which is clearly outlined by the ^{18}O contours. Because this fault is intimately connected with the quartz vein and the alteration pattern, it must have existed during the Miocene hydrothermal activity, but fault movement obviously continued subsequent to this activity. The offset of the ^{18}O contours indicates that the postalteration vertical displacement along the Comstock fault is between 500 and 1000 m, in agreement with independent geologic estimates (Criss and Champion, 1991). The great offset along this 45°, east-dipping fault has down-dropped a high-^{18}O tongue of rocks into the anomaly, and thereby preserved many of the famous, hanging-wall ore deposits from erosion.

Much additional information may be gleaned by examining vertical isotopic relationships in this area. In Figure 5.12, the δ^{18}O values of 75 rock specimens, mostly representing historical samples from the Con Virginia and California mines, are projected and contoured into the plane of the geologic cross section drawn by Becker (in 1882). This data base allows construction of a detailed image of the convective fluid flow that produced the Con Virginia orebody, better known as the "Big Bonanza." Figure 5.12 clearly shows the intricate and conformable relationship of the bonanza to a large gyre in the δ^{18}O contours. Meteoric-hydrothermal fluid originated in the cooler, surrounding country rocks, moved downward and laterally inward toward the Davidson granodiorite intrusion, and then ascended along and above the Comstock fault and the contact zone of the intrusion, in a pattern consistent with buoyancy-driven groundwater flow. The flow then appears to have overturned, moving downward and to the

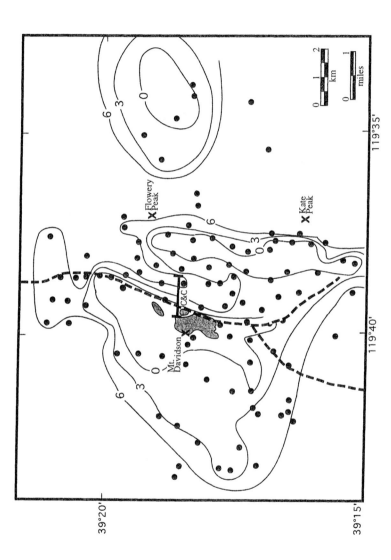

Figure 5.11 Contour map of whole-rock $\delta^{18}O$ values (circles) for andesites and granodiorites in the Virginia City area, Nevada. A circular, 75-km^2 zone of alteration is centered on the Mt. Davidson granodiorite stock (shaded area), which appears to represent the throat of an ancient stratovolcano. The isotopic anomaly has been disrupted by continued movement of the Comstock fault (dashed line), along which were formed some of the most spectacular Ag–Au orebodies in the world. The line "C&C" shows the location of the cross section through the famous "Big Bonanza" orebody shown in Figure 5.12. After Criss and Champion (1991).

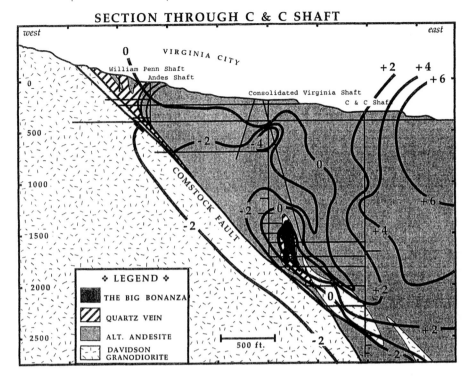

Figure 5.12 Vertical section through the "Big Bonanza" orebody, as drawn by Becker (in 1882), on which are superimposed contours for 75 whole-rock $\delta^{18}O$ values for altered andesites and granodiorites (see "C&C" in Figure 5.11). This famous orebody occupies an anomalous position above, rather than along, the Comstock fault. Patterns of alteration defined by the contours indicate that the Big Bonanza formed in a deep gyre of meteoric-hydrothermal fluid, convecting in a complex flow field consistent with the patterns discussed by Criss and Hofmeister (1991). The conformable relation between the orebody and the ^{18}O contours will help define drilling targets in this area. From Eaton et al. (1998).

east of the Big Bonanza orebody. The core of this gyre ($\delta^{18}O = 0$ to $+3.8$) encompasses the bonanza and is almost totally surrounded by a zone of extremely high ^{18}O gradients; this extends into a zone of very low $\delta^{18}O$ values (< -2.0) that represents zones dominated by hot, deep-circulating groundwater. Many features of this complex flow field are consistent with experimentally observed patterns of fluid convection in permeable media (Criss and Hofmeister, 1991).

5.4.4.2 Large Calderas and Ring Anomalies   The greatest volcanic eruptions on Earth are associated with giant calderas. These systems typically form when granitic batholiths intrude close to Earth's surface, and enormous, circular blocks of crustal rocks subside into the underlying magma chamber, displacing hundreds and sometimes even thousands of cubic kilometers of magma as ash-flow tuffs. While no historical eruptions have produced more than a tiny fraction of such volumes, the geologic record indicates that at least one enormous caldera has formed every million years in the western United States during the Tertiary period.

All the necessary ingredients for meteoric-hydrothermal systems are present in subaerial caldera systems, in that epizonal magma bodies are juxtaposed with highly fractured, permeable rocks along the ring faults. Indeed, rocks with very low $\delta^{18}O$ values have been encountered by drilling into the calderas at Yellowstone, Wyoming and at Long Valley, California, both of which formed during the last million years. Studies of older, more deeply eroded calderas show that the rocks with the lowest $\delta^{18}O$ values occur in annular zones that coincide with the ring zones, as documented in the Idaho batholith and at Lake City, Colorado (Criss and Taylor, 1983; Larson and Taylor, 1986).

Besides producing the striking, ring-shaped isotopic anomalies, the meteoric-hydrothermal alteration in caldera systems commonly produces regional propylitic alteration and distinctive, steep, disequilibrium trends on δ–δ plots. While the latter trends are clear indicators of subsolidus alteration in open systems, of special interest is the less common occurrence of unit-sloped, equilibrium trends on the same diagrams. The latter trends provide clear-cut evidence for low-^{18}O *magmas* in certain systems, as shown by the ash-flow tuffs on Figure 5.7. Taylor (1977) has persuasively argued that such unusual magmas can be produced by the stoping of huge blocks of altered, low-^{18}O country rocks during cauldron subsidence.

5.4.4.3 High-Temperature Hydrothermal Systems   Growing evidence shows that meteoric-hydrothermal alteration can occur under very-high-temperature conditions. As an example, the Skaergaard intrusion, east Greenland, is a 10 km × 7 km intrusion of tholeiitic magma that formed approximately 55 Ma ago during the opening of the North Atlantic Ocean. Deep glacial dissection of this pluton have exposed beautiful igneous layering, mostly formed from the bottom up, but also from the top down, and meeting at the "sandwich horizon" during the latest stage of crystallization. Large depletions in $\delta^{18}O$ values occur throughout the upper two-thirds of the stock, indicating the operation of a pervasive meteoric-hydrothermal system in the pluton that had been long regarded as the classical example of closed system cooling and fractional crystallization!

That the Skaergaard hydrothermal system began very early, even while the pluton was still crystallizing, is shown by many interesting isotopic features found by Taylor and Forester (1979). For example, large, low-^{18}O xenoliths are enclosed in less altered rock, and clearly represent highly altered blocks of roof rock that were assimilated by, and ultimately frozen within, the crystallizing magma. Norton and Taylor (1979) used a detailed, convective cooling model of the Skaergaard intrusion to conclude that most of the hydrothermal fluid passed through the pluton at temperatures exceeding 480°C. Such high temperatures were required not only to match the detailed $\delta^{18}O$ patterns mapped by Taylor and Forester (1979), but also to reconcile the large ^{18}O depletions with the general lack of the secondary hydrous minerals, such as chlorite, that would have formed had the meteoric-hydrothermal alteration occurred at lower temperatures. Such high temperatures are confirmed by the observations of Bird et al. (1988) that hydrothermal veins in this intrusion contain calcic amphibole and clinopyroxene. Such assemblages indicate vein formation at temperatures of 600–900°C, well into the "upper amphibolite" facies.

Oxygen isotope data also provide evidence for a new type of high-temperature, meteoric-hydrothermal system associated with fumarolic vents of ash-flow tuffs. In the 2.8-Ma ash-flow sheets from the Chegum caldera in the Caucasus Mountains, Russia, and in the 0.76-Ma Bishop Tuff erupted from the Long Valley caldera, California, Gazis et al. (1996) and Holt and Taylor (1998) found a new effect, notably the occurrence of low-$\delta^{18}O$ whole-rock values in tuffs that contain unaltered feldspars. On a δ–δ plot where the $\delta^{18}O$ values for the glassy groundmass are plotted against those for the feldspar, steep disequilibrium arrays are found that point to open system fluid infiltration. This effect is similar to the quartz–feldspar and plagioclase–pyroxene effects discussed previously, except, in this case, the feldspar is the slowly exchanging mineral, rather than the fastest! This is a key observation, because feldspars are normally very reactive and easily exchanged during hydrothermal activity. These relationships can be produced by very brief (10–25 years) interactions of the tuff with hot, 500–600°C, fumarolic fluid derived from meteoric water, under moderate water/rock ratios. During such brief events, normal rain falling directly on the tuff could contribute only a small fraction of the water required to explain the alteration patterns, so that lateral flow of meteoric groundwater through the permeable tuff units is seemingly required.

5.5 Extraterrestrial Oxygen and Hydrogen Isotopic Compositions

Stable isotopes provide some of the most important information on the origin and history of matter. These data establish a special relationship between Earth and the Moon, and also show, as a "rule of thumb," that isotopic differences among materials tend to increase with their increasing separation, both in space and time, in the solar system. Most peculiar of all are certain isotopic anomalies established in the earliest history of the solar system, and even before.

5.5.1 Oxygen Isotopes

5.5.1.1 Lunar Rocks The $\delta^{18}O$ values of lunar materials, including lithologies as diverse as mare basalts, breccias, and highland crystalline rocks, are remarkably uniform at +5.4 to +6.8. Lunar igneous rocks have an even smaller range of +5.7 ± 0.2 (Epstein and Taylor, 1971; Taylor and Epstein, 1973). These values reflect the simple geologic history of the Moon, its lack of water, and the fact that simple fractional crystallization processes can produce only small variations in the $\delta^{18}O$ values of significant volumes of melts during igneous differentiation, because at high temperatures the fractionation factors between silicates are small.

Minerals in lunar volcanic rocks exhibit a very close approach to isotopic equilibrium at a realistic extrusive temperature of about 1200°C. The order of ^{18}O enrichment among coexisting minerals is therefore that which would be expected from the high-temperature fractionation factors. "Typical" $\delta^{18}O$ values for lunar minerals are (Mayeda et al., 1975)

Cristobalite +6.8   Plagioclase +5.7   Pyroxene +5.4   Olivine +5.0

Ilmenite +4.0

Subsolidus effects, such as isotopic exchange during slow cooling, are minimal in lunar lavas, and the effects of hydrothermal alteration are negligible in these petrographically pristine rocks.

5.5.1.2 Earth–Moon System and the Three-Isotope Plot   The venerable idea that Earth and the Moon are closely related has become widely accepted since the *Apollo* missions, which returned lunar rocks with $\delta^{18}O$ values that were practically identical to those inferred for the bulk Earth. Even more compelling are combined $\delta^{18}O$ and $\delta^{17}O$ measurements that show that lunar and terrestrial materials lie along a common trend that is distinct from all other samples of extraterrestrial matter.

Figure 5.13 is a "three-isotope plot" in which $\delta^{17}O$ values are plotted against the $\delta^{18}O$ values of the same material (Clayton et al., 1976; Clayton, 1986). On such

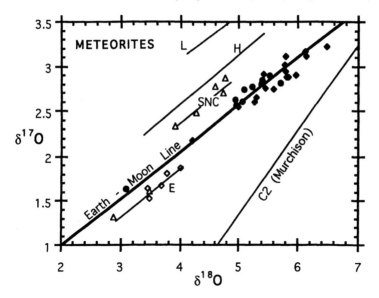

Figure 5.13 Graph of $\delta^{17}O$ versus $\delta^{18}O$ values for meteorites and representative lunar and terrestrial rocks. The bold "Earth–Moon" line represents terrestrial (filled circles) and lunar samples (filled diamonds), which together define a trend with a slope of about 0.52 that is consistent with their common bulk composition and their subsequent evolution by processes accompanied by ordinary isotopic fractionations. Different families of meteorites, such as the SNC achondrites, thought to come from Mars, and the eucrites (E), define similar slopes but have different bulk compositions than the Earth–Moon system. More primitive meteorites, such as the H and L chondrites, have still more divergent bulk compositions, and most unusual of all are trends for refractory inclusions within C2 and C3 chondrites, exemplified by the unit-sloped trend for the inclusions in the Murchison C2 meteorite. Mass-independent fractionations produced in the earliest condensates of the solar system are preserved in such meteorites. Data from Clayton and Mayeda (1975, 1983) and Clayton et al. (1976).

a plot, a uniform reservoir that has some given composition is represented by a single point. If all or part of that reservoir is subsequently divided into parts that differ in isotopic composition due to the operation of any normal, mass-dependent fractionation process, those parts will lie along a line that has a slope of 0.529 (equation 2.74). In other words, under the stated conditions, the system and all its derivative components are forever constrained to lie along the line of this slope that passes through the initial bulk composition of the reservoir.

On a three-isotope plot, all analyzed samples from Earth and the Moon lie along a single trend called the "Earth–Moon Line" (Figure 5.13). All terrestrial rocks fit this line, even going far back into the Precambrian period, so there is no evidence for any postaccretion heterogeneities in oxygen isotopes. Moreover, no other samples, such as meteorites, lie on the Earth–Moon line. This remarkable relationship establishes that Earth and the Moon formed in proximal parts of the solar nebula, and might even represent parts of a single precursor body.

5.5.1.3 Tektites and Glass Spherules   Tektites are typically small (a few centimeters), glassy rocks that occur in strewn fields that commonly are spatially associated with large terrestrial meteor craters. Many tektites are aerodynamically shaped, suggesting that they were partially molten during their high-velocity passage through the atmosphere. Oxygen isotopes provide definitive evidence that tektites represent Earth material that was melted upon impact, then ejected high into the atmosphere, ultimately to fall back. In particular, the $\delta^{18}O$ values of tektites are typically +8.9 to +11.5, and can be as high as +14.9 (Taylor and Epstein, 1969). Such high values are inconsistent with the popular idea of a lunar origin for tektites.

Glass spherules are similar to tektites, but much smaller. Blum and Chamberlain (1992) describe spherules from Haiti that are arguably related to a giant impact associated with the Cretaceous–Tertiary boundary. These black to yellow spherules have $\delta^{18}O$ values of +6.0 to +14.0 that are inversely correlated with their SiO_2 contents. The spherules must likewise represent impact melts of assorted terrestrial target rocks that appear to include a high-^{18}O carbonate endmember.

5.5.1.4 Meteorites   While only a few thousand meteorites have been found, these fascinating objects, most of which are representatives of the asteroid belt, have been the subject of extensive petrographic and geochemical study. Of particular interest here are the stony meteorites, which contrast with the "irons" in that they are primarily constituted of silicates, and so are amenable to oxygen isotope analysis. Stony meteorites are subdivided into the achondrites and the chondrites, depending respectively on the absence or presence of spheroidal silicate inclusions called chondrules (e.g., Henderson, 1982). Most meteorites retain geochronologic parent–daughter systems that are indicative of their formation about 4.6 Ga ago, during the condensation of the solar system. Moreover, the bulk composition of the most primitive chondritic meteorites provides the best estimates for the bulk composition of Earth, while the chemically differentiated meteorites—such as the achondrites and irons—provide invaluable information on the timescale and consequences of the internal differentiation of planets.

Achondrites are differentiated, chemically fractionated rocks that are petrographically similar to igneous rocks or breccias. On a three-isotope plot (Figure 5.13), the achondrites lie within about 1‰ of the Earth–Moon line (Clayton and Mayeda, 1983). In detail, the achondrites define two different trends that must represent two distinct reservoirs, arguably representing different parent bodies that accreted in different parts of the solar system. The first group includes the eucrites, howardites, diogenites, and certain stony–iron meteorites. The second group, known by the acronym "SNCs," arguably came from Mars, in that they exhibit young and distinctive crystallization ages of 1.4–0.65 AE, and have a high oxidation state and high volatile contents.

Chondrites are the most common meteorites. They are unlike any familiar rocks in that they have undergone little or no chemical differentiation, and contain a peculiar mix of constituents. The most primitive types, the carbonaceous chondrites, are composed of both high-temperature and low-temperature materials that appear to have been mechanically assembled during the formation of the solar system, and been little affected since. The high-temperature materials include silicate materials such as olivine, pyroxene, and so on; silicate inclusions such as the chondrules that might represent molten drops, along with blebs of metal and sulfide; and refractory, Ca and Al-rich inclusions known as CAIs. The low-temperature materials occur in a fine-grained matrix that includes hydrous materials such as phyllosilicates (e.g., chlorite), water-soluble salts such as magnesium sulfate, and volatile organic material. Excepting hydrogen, helium, and inert gases, the abundance of the elements in carbonaceous chondrites is practically identical to their relative abundances in the Sun.

On a three-isotope plot (Figure 5.13), the chondrites define trends that plot well outside the narrow achondrite band. On the low-^{18}O side are C1 chondrites and ordinary chondrites, while the high-^{18}O side contains the C2 and the C3 carbonaceous chondrites. The details of meteorite classification are of little importance here, the primary point being that oxygen isotope data have confirmed many of the petrographic assignments of meteorites to various families and inferred parent bodies.

Of special interest are relationships in primitive, C2 and C3 carbonaceous chondrites such as Murchison and Allende, whose dramatic fall was observed in Mexico in 1969. On the three-isotope plot, the low-temperature matrix materials of these meteorites fall along a slope 1/2 line. However, many high-temperature anhydrous minerals—such as olivine and pyroxene, and the CAIs—define a trend that has a unit slope (Figure 5.13; Clayton et al., 1973). Such a unit-sloped trend must project to the point that represents pure ^{16}O (possibly representing a pure nucleosynthetic component in the solar nebula) that condensed before it became homogenized. Such heterogeneities could have been produced in a supernova event that triggered the formation of the solar system, and different proportions of this component in different regions could have resulted in the different reservoirs of material that are defined by the various trends on the three-isotope plot (Clayton et al., 1973). Another possibility is that some type of mass-independent fractionation processes operated in the solar nebula.

226 Principles of Stable Isotope Distribution

5.5.2 Hydrogen Isotopes

Hydrogen isotope ratios vary enormously in meteorites, and even more widely among the bodies in the solar system. Even greater variations occur throughout the Galaxy. In part, this is because deuterium was produced in the Big Bang, but is destroyed by the nuclear reactions in stars. For example, Epstein and Taylor (1971) estimate that the D/H ratio of the solar wind is less than 3×10^{-6}, indicating that the Sun has completely burned up its original inventory of D. It has been argued that the D/H ratio of the interstellar medium is inversely related to time.

Table 5.3 provides measurements and estimates of D/H ratios for the Sun, several planets, and interstellar clouds. The values for Jupiter probably best represent the original value for the primitive solar nebula at the time of its origin, 4.55 Ga ago. Subsequently, the Sun has exhausted its deuterium, while Earth and especially Venus have become highly enriched in deuterium, probably due to loss of protium to space following the photochemical destruction of atmospheric water. Truly enormous values have been estimated for interstellar molecules in the dark clouds in the Orion nebula.

5.5.3 Anomalies in Other Stable Isotopes

Following the discovery of the anomalous oxygen isotope ratios in CAIs, isotope anomalies in numerous other elements, including Ca, Ti, Fe, Sr, Ba, and so on, were identified (see Zinner, 1989, 1998). These anomalies reflect the production of anomalous amounts of certain isotopes by nuclear processes. The anomalous compositions would have been eliminated had they not been rapidly incorporated within early condensates of the solar nebula, before homogenization of matter was complete.

In some cases, the anomalous concentrations of certain isotopes relate not to their direct nucleosynthesis, but to the nuclear production of their short-lived parent isotopes. Best known of these effects are the ^{26}Mg anomalies produced by the decay of extinct ^{26}Al. Normal aluminum contains solely the stable isotope ^{27}Al, but ^{26}Al was produced in the nucleosynthetic event that triggered the formation of the solar system. Subsequent decay of the ^{26}Al to its stable radiogenic daughter produced anomalous $^{26}Mg/^{24}Mg$ ratios that are preserved in refractory

Table 5.3 D/H Ratios of Astronomical Objects

Object	D/H (ppm)	Reference
Sun	< 3	Epstein and Taylor, 1971
Venus	~15,000	Baugher, 1988
Earth	150	Table 1.2
Jupiter	23; 51	Geiss and Reeves, 1981
	26	Fegley and Prinn, 1988
Saturn	17	Ditto
	59	Geiss and Reeves, 1981
Uranus	66	Ditto
Galaxy	20	Ditto
Orion nebula	1000–50,000	Ditto

inclusions in meteorites. Because the half-life of ^{26}Al is only 730,000 years, the interval between nucleosynthesis and the condensation of the refractory materials must have been very short—no longer than a few million years.

Even more fascinating is the existence of presolar grains, literally "stardust," in primitive meteorites. Compared with the isotopic anomalies in refractory inclusions that typically are on the order of a few per cent, stable isotope variations of Si, C, N, and O in presolar dust grains are huge, with the ratios commonly extending over several orders of magnitude (Zinner, 1996, 1998). The grains of SiC in chondrites carry such anomalies, and are believed to have formed in Red Giant stars that existed long before the solar system formed. Other anomalies have been identified in presolar diamond, graphite, corundum, and silicon nitride. The ion probe is remarkably well suited to study the large isotope variations in these micron-sized dust particles.

5.6 Summary

Oxygen and hydrogen isotope ratios provide remarkable information on the origin of rocks and on the processes that subsequently affect them. Lunar rocks, mantle rocks, and many igneous rocks derived from the mantle have highly uniform oxygen isotope ratios that are close to the bulk compositions of Earth and the Moon. Lunar igneous minerals have a simple history and so retain fractionation patterns produced during high-temperature crystallization, while complex subsolidus processes—such as slow cooling and hydrothermal alteration—have altered the isotopic patterns in igneous rocks on Earth. Even larger variations are produced by weathering and chemical processes during the generation of sedimentary and metamorphic rocks in the complex rock cycle on Earth. The combination of isotopic fractionation effects with material balance constraints provides a powerful means to investigate these processes.

Profound isotopic differences have been discovered in meteorites, and even among different inclusions within single primitive meteorites. Many of these differences record isotopic heterogeneities in the early solar system that were produced by a nucleosynthetic event, such as a supernova. Truly enormous isotopic differences have been found in minuscule dust inclusions that probably represent relicts of stars that existed before the solar system formed.

5.7 Problems

1. Suppose that $\alpha_{xl\text{-melt}}$ varies linearly with f during fractional crystallization of a melt, from a value of 0.998 initially to 1.002 at the end stage. Integrate the Rayleigh equation for this condition, and construct a graph of $\delta^{18}O$ versus f, assuming that the initial magma was +6.0‰.

2. A carbonatite magma, containing 50% molten $CaCO_3$ with a $\delta^{18}O$ value of +7.0 and a $\delta^{13}C$ value of −5.5, assimilates limestone country rock with a $\delta^{18}O$

value of $+25.0$ and a $\delta^{13}C$ value of $+1.0$. Assume that no CO_2 gas is released during this process, that 5 g of cumulate crystals form for every gram of limestone dissolved, and that the crystal–melt fractionation factor is 1.001 for oxygen and 1.000 for carbon. Construct a graph of the $\delta^{13}C$ versus the $\delta^{18}O$ values of the evolving magma and the crystalline rock formed from it. Place tick marks along the curve for every 10% crystallized.

3. In addition to the conditions stated in problem 2, the carbonatite magma has 6000 ppm Sr with an $^{87}Sr/^{86}Sr$ ratio of 0.7032, while the limestone has 600 ppm Sr with a ratio of 0.7090. Assuming that the strontium partition factor (Dsr) is 1.2, construct a graph of $^{87}Sr/^{86}Sr$ versus the inverse Sr concentration of the evolving magma, and also plot a graph of the $^{87}Sr/^{86}Sr$ values versus the $\delta^{18}O$ values.

4. Show that equation 5.14 reduces to the relation $\partial R_B/\partial R_A = -X_A/X_B$, for a two-phase system.

5. A gabbro with a whole-rock $\delta^{18}O$ value of $+5.7$—composed of 50 mol. % of oxygen in plagioclase (anorthite, 67:albite, 33), 35% clinopyroxene, 10% olivine, and 5% magnetite—cools slowly from its crystallization temperature of $900°C$. Plot a graph of the $\delta^{18}O$ values of the plagioclase versus those of the magnetite during cooling, assuming that (1) isotopic equilibrium is maintained among all the phases; and (2) isotopic equilibrium among the phases is maintained down to $500°C$, after which the magnetite, in effect, becomes isotopically isolated from the system because of slow exchange rates. In your calculations, use the following coefficients for "A" in the quartz–mineral fractionation factors given by Chiba et al. (1989): quartz–albite (0.94); quartz–anorthite (1.99); quartz–diopside (2.75); quartz–forsterite (3.67); and quartz–magnetite (6.29); where $\ln \alpha_{qtz\text{-}mineral} = 1000A/T^2$.

6. In four separate experimental runs conducted by Melchiorre (1998), 300 mg of powdered malachite and 35 ml of various waters were sealed in small bottles and left to react at $25°C$. After 216 days, the capsules were analyzed with the results tabulated below.

Run	$\delta^{18}O_{w,initial}$	$\delta^{18}O_{mal,initial}$	$\delta^{18}O_{w,final}$	$\delta^{18}O_{mal,final}$
1	$+50.2$	$+25.61$	$+49.1$	$+31.08$
2	$+101.3$	$+25.61$	$+99.9$	$+36.46$
3	-34.80	$+25.61$	-34.80	$+22.26$
4	-74.00	$+25.61$	-73.70	$+17.80$

Calculate the mineral/water ratios, X_{mal}/X_w, for each of these runs. Why might these values differ from the charge loaded in the experiments? Also, use the partial exchange technique to estimate the equilibrium fractionation factor and the fraction of exchange, that is, plot α_i versus $\alpha_f - \alpha_i$. Then, use the corrected technique to refine these values. How long would be required for the exchange process to reach 90% completion?

7. Craft an approximate oxygen isotope fractionation equation for the biotite–water system, utilizing the albite–water and anorthite–water equations of O'Neil

and Taylor (1967; see Appendix A.4), together with the determination of Bottinga and Javoy (1975) that the fractionation between An60 and biotite has values of -0.60 for c_1 and 2.096 for c_3, again in the form of the equation in Appendix A.4. What assumptions are necessary? Is biotite enriched or depleted in ^{18}O relative to water at 300°C, and by how much?

8. An altered Tertiary granite analyzed by Criss and Taylor (1983; sample RH 120) has approximately 30% quartz with a $\delta^{18}O$ value of $+10.1$, 65% alkali feldspar with a $\delta^{18}O$ value of $+6.2$, and 5% biotite with a $\delta^{18}O$ value of $+0.2$ and a δD value of -150. Assuming that the quartz originally had a $\delta^{18}O$ value of $+10.3$, estimate the $\delta^{18}O$ values of the feldspar, biotite, and whole rock when the granite initially crystallized. Next, assuming that the granite underwent hydrothermal alteration at 300°C with fluid having an initial $\delta^{18}O$ value of -15.0, calculate the closed and open system $\mathcal{W}/\mathcal{R}$ ratios for each of the minerals, and for the calculated whole rock. What assumptions are necessary? Why do the estimated $\mathcal{W}/\mathcal{R}$ ratios differ?

9. In a high-latitude region, heated (300°C) meteoric water with a $\delta^{18}O$ value of $-14.0‰$ circulates through a granodiorite stock with $\delta^{18}O$ values of $+10.0$ for quartz, $+8.5$ for alkali feldspar, and $+5.0$ for biotite. If this process continues indefinitely, what $\delta^{18}O$ values will these minerals ultimately attain? Assuming that the fluid flow rate is very rapid compared with the isotopic exchange rate, and that the feldspar and biotite exchange 10× faster than the quartz, on a δ–δ plot graph the trajectory followed by the minerals.

References

Albarede, F. (1995) *Introduction to Geochemical Modeling.* Cambridge University Press, Cambridge.

Anderson, T.F. and Arthur, M.A. (1983) Stable isotopes of oxygen and carbon and their application to sedimentologic and paleoenvironmental problems. In *Stable Isotopes in Sedimentary Geology, SEPM Short Course,* Vol. 10, Society of Economic Paleontologists and Mineralogists, pp. 1-1–1-151.

Baugher, J.F. (1988) *The Space-Age Solar System.* John Wiley & Sons, New York.

Bird, D.K., Manning, C.E., and Rose, N.M. (1988) Hydrothermal alteration of Tertiary layered gabbros, east Greenland. *Am. J. Sci.,* **288,** 405–457.

Blum, J.D. and Chamberlain, C.P. (1992) Oxygen isotope constraints on the origin of impact glasses from the Cretaceous–Tertiary boundary. *Science,* **257,** 1104–1107.

Bottinga, Y. and Javoy, M. (1975) Oxygen isotope partitioning among the minerals in igneous and metamorphic rocks. *Rev. Geophys. Space Phys.,* **13,** 401–418.

Chiba, H.T., Chacko, T., Clayton, R.N., and Goldsmith, J.R. (1989) Oxygen isotope fractionation involving diopside, forsterite, magnetite, and calcite: application to geochemistry, *Geochim. Cosmochim. Acta,* **53,** 2985–2995.

Clayton, R.N. (1986) High temperature isotope effects in the early solar system, *Rev. Mineral.,* **16,** 129–140.

Clayton, R.N. and Mayeda, T.K. (1975) Genetic relations between the moon and meteorites. *Proc. 6th Lunar Sci. Conf.,* **6,** 1761–1769.

Clayton, R.N. and Mayeda, T.K. (1983) Oxygen isotopes in eucrites, shergottotes, nakhlites, and chassignites, *Earth Planet. Sci. Lett.,* **62,** 1–6.

Clayton, R.N., O'Neil, J.R. and Mayeda, T.K. (1972) Oxygen isotope fractionation between quartz and water, *J. Geophys. Res.*, **77**, 3057–3067.

Clayton, R.N., Grossman, L., and Mayeda, T.K. (1973) A component of primitive nuclear composition in carbonaceous meteorites. *Science*, **182**, 485–488.

Clayton, R.N., Onuma, N., and Mayeda, T.K. (1976) A classification of meteorites based on oxygen isotopes, *Earth Planet. Sci. Lett.*, **30**, 10–18.

Cortecci, G., Del Moro, A., Leone, G., and Pardini, G.C. (1979) Correlation between strontium and oxygen isotopic compositions of rocks from the Adamello massif (Northern Italy). *Contrib. Mineral. Petrol.*, **68**, 421–428.

Crank, J. (1975) *The Mathematics of Diffusion*. Oxford University Press, Oxford.

Criss, R.E. (1995) Stable isotope distribution: variations from temperature, organic and water–rock interactions. In *Global Earth Physcis: A Handbook of Physical Constants*, Ahrens, T.J., ed. AGU Reference Shelf 1, AGU, Washington, D.C., pp. 292–307.

Criss, R.E. and Champion, D.E. (1991) Oxygen isotope study of the fossil hydrothermal system in the Comstock Lode mining district, Nevada. *Geochem. Soc. Spec. Pub.*, **3**, 437–447.

Criss, R.E. and Fleck, R.J. (1990) Oxygen isotope map of the giant metamorphic-hydrothermal system around the northern part of the Idaho batholith, U.S.A. *Appl. Geochem.*, **5**, 641–655.

Criss, R.E. and Gregory, R.T. (1990) Equation for the slope of closed-system trends on the δ–δ and δ–Δ diagrams of stable isotope geochemistry. *Geol. Soc. America Abstr. Prgms*, **22**, A251.

Criss, R. E. and Hofmeister, A.M. (1991) Application of fluid dynamics principles in tilted permeable media to terrestrial hydrothermal systems. *Geophys. Res. Lett.*, **18**(2), 199–202.

Criss, R.E. and Taylor, H.P., Jr. (1983) An $^{18}O/^{16}O$ and D/H study of Tertiary hydrothermal systems in the southern half of the Idaho batholith, *Bull. Geol. Soc. America*, **94**, 640–663.

Criss, R.E. and Taylor, H.P., Jr. (1986) Meteoric-hydrothermal systems, *Rev. Mineral.*, **16**, 373–424.

Criss, R.E., Gregory, R.T., and Taylor, H.P., Jr. (1987) Kinetic theory of oxygen isotopic exchange between minerals and water, *Geochim. Cosmochim. Acta*, **51**, 1099–1108.

DePaolo, D.J. (1981) Trace element and isotopic effects of combined wallrock assimilation and fractional crystallization. *Earth Planet. Sci. Lett.*, **53**, 189–202.

Eaton, G.F., Criss, R.E., and Champion, D.E. (1998) The Becker Collection, Comstock Lode, Nevada: a compilation of historical notes and recent data. Lawrence Livermore National Laboratory Open-File Report, UCRL-ID-130195.

Epstein, S. and Taylor, H.P. Jr. (1971) O^{18}/O^{16}, Si^{30}/Si^{28}, D/H and C^{13}/C^{12} ratios in lunar samples, *Proc. 2nd Lunar Sci. Conf.*, **2**, 1421–1441.

Fegley, B. and Prinn, R.G. (1988) The predicted abundances of deuterium-bearing gases in the atmospheres of Jupiter and Saturn. *Astrophys. J.*, **326**, 490–508.

Fleck, R.J. and Criss, R.E. (1985) Strontium and oxygen isotopic variations in Mesozoic and Tertiary plutons of central Idaho. *Contrib. Mineral. Petrol.*, **90**, 291–308.

Friedman, I. and O'Neil, J.R. (1977) Compilation of stable isotope fractionation factors of geochemical interest. In *Data of Geochemistry*, Fleischer, M., ed. U.S. Geol. Survey Professional Paper, 440 KK.

Garlick, G.D. and Epstein, S. (1967) Oxygen isotope ratios in coexisting minerals of regionally metamorphosed rocks. *Geochim. Cosmochim. Acta*, **31**, 181–214.

Gazis, C., Taylor, H.P., Jr., Hon, K., and Tsvetkov, A. (1996) Oxygen isotopic and geochemical evidence for a short-lived, high-temperature hydrothermal event in the Chegem caldera, Caucasus Mountains, Russia. *J. Volcanol. Geotherm. Res.*, **73**, 213–244.

Geiss, J. and Reeves, H. (1981) Deuterium in the solar system. *Astron. Astrophys.*, **93**, 189–199.

Giletti, B. (1986) Diffusion effects on oxygen isotope temperatures of slowly cooled igneous and metamorphic rocks. *Earth Planet. Sci. Lett.*, **77**, 218–228.

Gregory, R.T. and Criss, R.E. (1986) Isotopic exchange in open and closed systems. *Rev. Mineral.*, **16**, 91–128.

Gregory, R.T. and Taylor, H.P., Jr. (1981) An oxygen isotope profile in a section of Cretaceous oceanic crust, Samail ophiolite, Oman: evidence for $\delta^{18}O$-buffering of the oceans by deep (> 5 km) seawater-hydrothermal circulation at mid-ocean ridges. *J. Geophys. Res.*, **86**, 2737–2755.

Gregory, R.T., Criss, R.E., and Taylor, H.P., Jr. (1989) Oxygen isotope exchange kinetics of mineral pairs in closed and open systems: applications to problems of hydrothermal alteration of igneous rocks and Precambrian iron formations. *Chem. Geol.*, **75**, 1–42.

Henderson, P. (1982) *Inorganic Geochemistry.* Pergamon Press, New York.

Holt, E.W. and Taylor, H.P., Jr. (1998) $^{18}O/^{16}O$ mapping and hydrogeology of a short-lived (~10 years) fumarolic (> 500°C) meteoric-hydrothermal event in the upper part of the 0.76 Ma Bishop tuff outflow sheet, California. *J. Volcanol. Geotherm. Res.*, **83**, 115–139.

Kamilli, R.J. and Criss, R.E. (1996) Genesis of the Silsilah tin deposit, Kingdom of Saudi Arabia. *Econ. Geol.*, **91**, 1414–1434.

Keith, M.L. and Weber, J.N. (1964) Carbon and oxygen isotopic composition of selected limestones and fossils. *Geochim. Cosmochim. Acta*, **28**, 1787–1816.

Knauth, L.P. and Epstein, S. (1976) Hydrogen and oxygen isotope ratios in nodular and bedded cherts, *Geochim. Cosmochim. Acta*, **40**, 1095–1108..

Kyser, T.K. (1986) Stable isotopic variations in the mantle. *Rev. Mineral.*, **16**, 141–164.

Larson, P.B. and Taylor, H.P., Jr. (1986) An oxygen isotope study of hydrothermal alteration in the Lake City caldera, San Juan Mountains, Colorado. *J. Volcanol. Geotherm. Res.*, **30**, 47–82.

Mayeda, T.K., Shearer, J., and Clayton, R.N. (1975) Oxygen isotope fractionation in Apollo 17 rocks. Proc. 6th Lunar Sci. Conf., pp. 1799–1802.

Melchiorre, E.B. (1998) Geochemical Studies of Low-Temperature Fluids in Diverse Geologic Settings. Unpublished Ph.D. thesis, Washington University, St. Louis, Missouri.

Muehlenbachs, K. and Clayton, R.N. (1976) Oxygen isotope composition of the oceanic crust and its bearing on seawater. *J. Geophys. Res.*, **81**, 4365–4369.

Muehlenbachs, K. and Kushiro, I. (1974) Oxygen isotopic exchange and equilibrium of silicates with CO_2 or O_2. *Carnegie Inst. Washington Yearbook*, **73**, 232–236.

Nabelek, P.I., Labotka, T.C., O'Neil, J.R., and Papike, J.J. (1984) Contrasting fluid/rock interaction between the Notch Peak granitic intrusion and argillites and limestones in western Utah: evidence from stable isotopes and phase assemblages. *Contrib. Mineral. Petrol.*, **86**, 25–34.

Northrop, D.A. and Clayton, R.N. (1966) Oxygen-isotope fractionations in systems containing dolomite. *J. Geol.*, **74**, 174–196.

Norton, D. and Taylor, H.P., Jr. (1979) Quantitative simulation of the hydrothermal systems of crystallizing magmas on the basis of transport theory and oxygen isotope data: an analysis of the Skaergaard intrusion. *J. Petrol.*, **20**, 421–486.

Rose, T.P., Criss, R.E., Mughannam, A.J., and Clynne, M.A. (1994) Oxygen isotope evidence for hydrothermal alteration within a Quaternary stratovolcano, Lassen Volcanic National Park, California. *J. Geophys. Res.*, **99**, 21,621–21,633.

Rye, R.O., Schuiling, R.D., Rye, D.M., and Janssen, J.B.H. (1976) Carbon, hydrogen, and oxygen isotope studies of the regional metamorphic complex at Naxos, Greece. *Geochim. Cosmochim. Acta*, **40**, 1031–1049.

Savin, S.M. and Epstein, S. (1970c) The oxygen isotope compositions of coarse grained sedimentary rocks and minerals. *Geochim. Cosmochim. Acta*, **34**, 25–42.

Savin, S.M. and Epstein, S. (1970b) The oxygen and hydrogen isotope geochemistry of ocean sediments and shales. *Geochim. Cosmochim. Acta*, **34**, 323–329.

Savin, S.M. and Epstein, S. (1970a) The oxygen and hydrogen isotope geochemistry of clay minerals. *Geochim. Cosmochim. Acta*, **34**, 25–42.

Sheppard, S.M.F. and Epstein, S. (1970) D/H and $^{18}O/^{16}O$ ratios of minerals of possible mantle or lower crustal origin. *Earth Planet. Sci. Lett.*, **9**, 232–239.

Shieh, Y.N. and Schwarcz, H.P. (1974) Oxygen isotope studies of granite and migmatite, Grenville province of Ontario, Canada. *Geochim. Cosmochim. Acta*, **38**, 21–45.

Stuckless, J.S. and O/Neil, J.R. (1973) Petrogenesis of the Superstition–Superior volcanic area as inferred from strontium- and oxygen-isotope studies. *Bull. Geol. Soc. America*, **84**, 1987–1998.

Suzuoki, T. and Epstein, S. (1976) Hydrogen isotope fractionation between OH-bearing minerals and water. *Geochim. Cosmochim. Acta*, **40**, 1229–1240.

Taylor, B.E. (1986) Magmatic volatiles: isotopic variations of C, H and S. *Rev. Mineral.*, **16**, 185–219.

Taylor, H.P., Jr. (1968) The oxygen isotope geochemistry of igneous rocks. *Contrib. Mineral. Petrol.*, **19**, 1–71.

Taylor, H.P., Jr. (1970) Oxygen isotope studies of anorthosites, with particular reference to the origin of bodies in the Adirondack Mountains, New York. In *Origin of Anorthosites*, New York State Museum Science Service Memoir 18, pp. 111–134.

Taylor, H.P., Jr. (1971) Oxygen isotope evidence for large-scale interaction between meteoric ground waters and Tertiary granodiorite intrusions, Western Cascade Range, Oregon. *J. Geophys. Res.*, **76**, 7855–7874.

Taylor, H.P., Jr. (1977) Water/rock interactions and the origin of H_2O in granitic batholiths. *J. Geol. Soc. London*, **133**, 509–558.

Taylor, H.P., Jr. (1980) The effects of assimilation of country rocks by magmas on $^{18}O/^{16}O$ and $^{87}Sr/^{86}Sr$ systematics in igneous rocks. *Earth Planet. Sci. Lett.*, **47**, 243–254.

Taylor, H.P., Jr. (1986) Igneous rocks. II. Isotopic case studies of circumpacific magmatism. *Rev. Mineral.*, **16**, 273–317.

Taylor, H.P., Jr. and Epstein, S. (1969) Correlations between O^{18}/O^{16} ratios and chemical compositions of tektites. *J. Geophys. Res.*, **74**, 6834–6844.

Taylor, H.P., Jr. and Epstein, S. (1973) O^{18}/O^{16} and Si^{30}/Si^{28} studies of some Apollo 15, 16, and 17 samples. *Proc. 4th Lunar Sci. Conf.*, **2**, 1657–1679.

Taylor, H.P., Jr. and Forester, R.W. (1979) An oxygen and hydrogen isotope study of the Skaergaard intrusion and its country rocks: A description of a 55-m.y. old fossil hydrothermal system. *J. Petrol.*, **20**, 355–419.

Taylor, H.P. and Sheppard, S.M.F. (1986) Igneous rocks. I. Processes of isotopic fractionation and isotope systematics. *Rev. Mineral.*, **16**, 227–271.

Valley, J.W. (1986) Stable isotope geochemistry of metamorphic rocks. *Rev. Mineral.*, **16**, 445–489.

Zinner, E. (1989) Isotopic measurements with the ion microprobe. In *New Frontiers in Stable Isotope Research*, Shanks, W.C. and Criss, R.E., eds. U.S. Geological Survey Bulletin, Vol. 1890, pp. 145–162.

Zinner, E. (1996) Presolar material in meteorites: an overview. In *Astrophysical Implications of the Laboratory Study of Presolar Materials*, Bernatowitz, T.J. and Zinner, E., eds, AIP Conf. Proceedings 402, American Institute of Physics, Woodbury, New York, pp. 3–26.

Zinner, E. (1998) Stellar nucleosynthesis and the isotopic composition of presolar grains from primitive meteorites. *Annu. Rev. Earth Planet. Sci.*, **26**, 147–188.

Appendix A.1

Important Nuclides of Light Elements

Element	Isotope	Mass (amu)	Abundance (atom %)	Half-life (years)
1. Hydrogen		**1.0079**		
	^{1}H	1.00782503	99.985	
	^{2}H	2.01410178	0.015	
	^{3}H			12.3
2. Helium		**4.002602**		
	^{3}He	3.01602930	0.000138	
	^{4}He	4.00260323	99.999862	
3. Lithium		**6.941**		
	^{6}Li	6.015122	7.5	
	^{7}Li	7.016003	92.5	
4. Beryllium		**9.012182**		
	^{9}Be	9.0121822	100.0	
	^{10}Be			1.6 E 6
5. Boron		**10.811**		
	^{10}B	10.0129372	19.9	
	^{11}B	11.0093056	80.1	
6. Carbon		**12.011**		
	^{12}C	12.0000000	98.90	
	^{13}C	13.00335483	1.10	
	^{14}C			5730.
7. Nitrogen		**14.0067**		
	^{14}N	14.00307400	99.63	
	^{15}N	15.00010896	0.37	
8. Oxygen		**15.9994**		
	^{16}O	15.99491462	99.76	
	^{17}O	16.9991314	0.04	
	^{18}O	17.999160	0.20	
9. Fluorine		**18.998403**		
	^{19}F	18.9984032	100.00	
10. Neon		**20.180**		
	^{20}Ne	19.992434	90.48	
	^{21}Ne	20.993841	0.27	
	^{22}Ne	21.991382	9.25	
11. Sodium		**22.98977**		
	^{23}Na	22.989767	100.00	
12. Magnesium		**24.305**		
	^{24}Mg	23.985042	78.99	
	^{25}Mg	24.985837	10.00	
	^{26}Mg	25.982594	11.01	
13. Aluminum		**26.981539**		
	^{26}Al			7.3 E 5
	^{27}Al	26.981538	100.00	
14. Silicon		**28.0855**		
	^{28}Si	27.976927	92.23	
	^{29}Si	28.976494	4.67	
	^{30}Si	29.973770	3.10	
	^{32}Si			100.
15. Phosphorus		**30.973762**		
	^{31}P	30.973762	100.00	

Element	Isotope	Mass (amu)	Abundance (atom %)	Half-life (years)
16. Sulfur		**32.07**		
	^{32}S	31.9720705	95.02	
	^{33}S	32.9714583	0.75	
	^{34}S	33.9678665	4.21	
	^{36}S	35.9670808	0.02	
17. Chlorine		**35.453**		
	^{35}Cl	34.96885272	75.77	
	^{36}Cl			3.01 E 5
	^{37}Cl	36.9659026	24.23	
18. Argon		**39.948**		
	^{36}Ar	35.967546	0.337	
	^{38}Ar	37.962732	0.063	
	^{39}Ar			269.
	^{40}Ar	39.962384	99.600	
19. Potassium		**39.0983**		
	^{39}K	38.963708	93.2581	
	^{40}K	39.964	0.0117	1.28 E 9
	^{41}K	40.961827	6.7302	
20. Calcium		**40.078**		
	^{40}Ca	39.962592	96.941	
	^{41}Ca			1.03 E 5
	^{42}Ca	41.958618	0.647	
	^{43}Ca	42.958767	0.135	
	^{44}Ca	43.955481	2.086	
	^{46}Ca	45.953687	0.004	
	^{48}Ca	47.952535	0.187	

Data from: Walker, F.W., Parrington, J.R., and Feiner, F. (1989) *Nuclides and Isotopes*, 14th ed., General Electric Co., San Jose, California.

Appendix A.2

Selected Physical Constants

Atomic mass unit, amu	$1.6605402 \times 10^{-27}$ kg
	931.49432 MeV
Avogadro constant	6.02214×10^{23} mol^{-1}
Boltzmann constant, k	1.380658×10^{-23} J/K
Coulomb force constant, $1/4\pi\varepsilon_0$	8.98755×10^{9} N-m^2/C^2
Elementary charge, q	1.6022×10^{-19} C
Gas constant, R	8.31451 J/mol-deg
Gas constant, R	1.9859 cal/mol-deg
Gas constant, R	0.08205 l-atm/mol-deg
Gas constant, R	8.31451 Pa m^3/mol-deg
Gravitational constant, G	6.6726×10^{-11} m^3/kg-s^2
Planck constant, h	6.62608×10^{-34} J-s
Rydberg constant, $\mathscr{R}$	1.097373×10^{7} m^{-1}
Speed of light, c	2.99792×10^{8} m/s

Data from: Lide, D.R. (1991) *Handbook of Chemistry and Physics*, 71st ed., CRC Press, Boston.

Appendix A.3

Definitions, Formulas, and Approximations

A.3.1 Definitions: Delta Value and Isotopic Fractionation Factor

$$\delta_A = 1000\left(\frac{R_A}{R_{std}} - 1\right) \tag{A.3.1.1}$$

$$\alpha_{A-B} = \frac{R_A}{R_B} = \frac{1000 + \delta_A}{1000 + \delta_B} \tag{A3.1.2}$$

A.3.2 Relationships between δ and R

$$\frac{R_A}{R_B} = \frac{1000 + \delta_A}{1000 + \delta_B} \tag{A.3.2.1}$$

$$\frac{R_A - R_B}{R_C - R_D} = \frac{\delta_A - \delta_B}{\delta_C - \delta_D} \tag{A3.2.2}$$

A.3.3 Differential Relationships between δ and R

$$\frac{dR_A}{dR_B} = \frac{d\delta_A}{d\delta_B} \tag{A3.3.1}$$

$$d\ln R_A = d\ln(1000 + \delta_A) = \frac{d\delta_A}{1000 + \delta_A} \tag{A.3.3.2}$$

For two elements, M and N:

$$\frac{(d\ln R_A)_{\text{element } M}}{(d\ln R_B)_{\text{element } N}} = \frac{d\ln(1000 + \delta_A)_{\text{element } M}}{d\ln(1000 + \delta_B)_{\text{element } N}} \tag{A.3.3.3}$$

A.3.4 Standard Conversion Identity

$$(\delta_A)_{\text{rel.std2}} = (\delta_A)_{\text{rel.std1}} + (\delta_{\text{std1}})_{\text{rel.std2}} + \frac{1}{1000}(\delta_A)_{\text{rel.std1}}(\delta_{\text{std1}})_{\text{rel.std2}} \quad \text{(A.3.4.1)}$$

or

$$\beta = \frac{(1000 + \delta_A)_{\text{rel.std\#2}}}{(1000 + \delta_A)_{\text{rel.std\#1}}} \quad \text{where the constant,} \quad \beta = \frac{1000 + (\delta_{\text{std1}})_{\text{rel.std\#2}}}{1000}$$

$$\text{(A.3.4.2)}$$

A.3.5 Material Balance Equations for Trace Isotope

$$R_{\text{system}} = \sum_{i=1}^{n} X_i R_i \quad \text{(A3.5.1)}$$

$$\delta_{\text{system}} = \sum_{i=1}^{n} X_i \delta_i \quad \text{(A.3.5.2)}$$

$$1 - F = \left(\frac{R_w^f}{R_w^i}\right)\left(\frac{\alpha_{A-W}^f - \alpha_{A-W}^{eq}}{\alpha_{A-W}^i - \alpha_{A-W}^{eq}}\right) \quad \text{(A.3.5.3)}$$

A.3.6 Mixing Equation for both Trace and Abundant Isotopes

$$\left[\left(\frac{C_B}{C_A}\right)\left(\frac{R_{\text{Mix}}^\dagger - R_B^\dagger}{R_{\text{Mix}}^\dagger - R_A^\dagger}\right)\right]_{\text{Element } L} = \left[\left(\frac{C_B}{C_A}\right)\left(\frac{R_{\text{Mix}}^\dagger - R_B^\dagger}{R_{\text{Mix}}^\dagger - R_A^\dagger}\right)\right]_{\text{Element } M}$$

$$= \cdots = -\frac{\text{Mass } A}{\text{Mass } B}$$

where

$$R^\dagger = \frac{\text{mass of isotope}}{\text{total mass of element}} \quad \text{(A.3.6)}$$

A.3.7 Process Differential Equations for Trace Isotopes

Rayleigh equation:

$$d\ln R = (\alpha - 1)\, d\ln f \quad \text{(A.3.7.1)}$$

Kinetic exchange

$$\frac{dR_A}{dt} = k_{A-B} B(R_A - \alpha_{A-B} R_B) \quad \text{(A.3.7.2)}$$

Evaporation of water into an atomosphere of humidity h:

$$dR_w = u(R_w - R_w^s)\, d\ln f \qquad\qquad (A.3.7.3)$$

where

$$R_w^s = \frac{\alpha_{eq} h R_v}{1 - \alpha_{evap}^0(1 - h)} \qquad \text{and} \qquad u = \frac{1 - \alpha_{evap}^0(1 - h)}{\alpha_{evap}^0(1 - h)}$$

A.3.8 Process Differential Equations for Abundant Minor Isotopes

Rayleigh equation

$$d\ln R = (\alpha - 1)\{d\ln f - d\ln(1 + R)\} \qquad\qquad (A.3.8.1)$$

Kinetic exchange

$$\frac{d\ln(1 + R_A)}{dt} = k_{A-B} B(R_A - \alpha_{A-B} R_B) \qquad\qquad (A.3.8.2)$$

A.3.9 Chemical Relationships

Relationship between R and the activities and symmetry numbers of isotopic species

$$R = \left(\frac{a^* \sigma^*}{a\sigma}\right)^{1/\text{diff}} \qquad\qquad (A.3.9.1)$$

where "diff" is the difference between the number of "heavy" isotopes in the molecular species.

Combinatorial Formula for Number of Distinct Isotopomers

$$\#\text{ Species} = \frac{(p + s - 1)!}{(p - 1)!\, s!} \qquad\qquad (A.3.9.2)$$

where $s = \#$ of identical sites and $p = \#$ of isotopes.

Limiting, high-temperature equilibrium constant for isotopic exchange reactions

$$K_\infty = \prod_i \sigma_i^{-b_i} \qquad\qquad (A.3.9.3)$$

where the b_i are stoichiometric coefficients.

Ferronsky and Polyakov rule

$$\alpha = \left(\frac{K}{K_\infty}\right)^{1/n} \qquad\qquad (A.3.9.4)$$

where n is the number of atoms exchanged.

Vapor pressure rule

$$(P^0_{AM_{b-a}M^*_a})^b = (P^0_{AM_b})^{b-a}(P^0_{AM^*_b})^a \tag{A.3.9.5}$$

A.3.10 Spectroscopic Equations for a Simple Harmonic Oscillator

Reduced mass of a diatomic molecule

$$\mu \equiv \frac{m_1 m_2}{m_1 + m_2} \tag{A.3.10.1}$$

SHO vibrational frequency relationship for diatomic molecules bearing isotopes m or m^* on a common substrate M

$$\left(\frac{\nu_{Mm^*}}{\nu_{Mm}}\right)^2 = \frac{\mu_{Mm}}{\mu_{Mm^*}} \tag{A.3.10.2}$$

SHO frequency relationship for homogeneous diatomic molecules and two isotopomers: for example, H_2, HD, and D_2

$$\nu^2_{mm^*} = \frac{\nu^2_{mm} + \nu^2_{m^*m^*}}{2} \tag{A.3.10.3}$$

SHO frequency relationship for heterogeneous diatomic molecules

$$\nu^2_{mm} - \nu^2_{Mm^*} = \nu^2_{Mm}\left(\frac{M}{m^*}\right)\left(\frac{m^* - m}{m + M}\right) \tag{A.3.10.4}$$

SHO frequency relationship for various isotopes on a common substrate M

$$\nu^2_{Mm^*} - \nu^2_{Mm} = z(\nu^2_{Mm^\#} - \nu^2_{Mm}) \qquad \text{where } z = \left(\frac{m^\#}{m^*}\right)\left(\frac{m^* - m}{m^\# - m}\right) \tag{A.3.10.5}$$

SHO Vibrational Energy Levels

$$E_n = (n + \tfrac{1}{2})h\nu \tag{A.3.10.6}$$

SHO Zero Point Energy

$$\text{ZPE} = \frac{h\nu}{2} \tag{A.3.10.7}$$

A.3.11 Derivative of Integrals

$$\frac{d}{dx}\int_{u(x)}^{v(x)} f_{(t)}\, dt = f_{[v(x)]}\frac{dv}{dx} - f_{[u(x)]}\frac{du}{dx} \tag{A.3.11.1}$$

A.3.12 Useful Mathematical Approximations

$$e^x \cong 1 + x \qquad \text{for small } x \qquad \text{(A.3.12.1)}$$
$$\ln(1 + x) \cong x \qquad \text{for small } x \qquad \text{(A.3.12.2)}$$
$$(1 \pm x)^n \cong 1 \pm nx \qquad \text{for small } x \qquad \text{(A.3.12.3)}$$
$$\ln x! \cong x \ln x - x \qquad \text{for large } x \qquad \text{(A.3.12.4)}$$

A.3.13 Fractionation Factor Approximations for $\alpha \sim 1$

Definition of Big Delta

$$\Delta_{A-B} \equiv \delta_A - \delta_B \qquad \text{(A.3.13.1)}$$

Related Approximations

$$\Delta_{A-B} \cong 1000(\alpha_{A-B} - 1) \qquad \text{(A.3.13.2)}$$
$$\Delta_{A-B} \cong 1000 \ln \alpha_{A-B} \qquad \text{(A.3.13.3)}$$
$$p\Delta_{A-B} \cong 1000(\alpha_{A-B}^p - 1) \qquad \text{where } p \text{ is a constant} \qquad \text{(A.3.13.4)}$$
$$p(\alpha_{A-B} - 1) \cong \alpha_{A-B}^p - 1 \qquad \text{where } p \text{ is a constant} \qquad \text{(A.3.13.5)}$$

A.3.14 Error Function

$$\operatorname{erf} u = \frac{2}{\sqrt{\pi}} \int_0^u e^{-\xi^2} d\xi \qquad \text{(A.3.14.1)}$$

$$\operatorname{erf}(0) = 0 \qquad \operatorname{erf}(\infty) = 1 \qquad \operatorname{erf}(-u) = -\operatorname{erf}(u) \qquad \text{(A.3.14.2)}$$

$$\operatorname{erf}(1) = 0.842701 \ldots \qquad \text{(A.3.14.3)}$$

$$\operatorname{erfc}(u) = 1 - \operatorname{erf}(u) \qquad \text{(A.3.14.4)}$$

Error function approximations

$$\operatorname{erf}(u) \cong \frac{2u}{\sqrt{\pi}} \qquad \text{for small } u \qquad \text{(A.3.14.5)}$$

$$\operatorname{erf}(u) \cong \sqrt{1 - \exp(-4u^2/\pi)} \qquad \text{(A.3.14.6)}$$

Appendix A.4

Selected Oxygen Isotope Fractionation Factors between Various Phases and Water

---

General form: $1000 \ln \alpha^{eq}_{A-\text{water}} = C_1 + 10^3 C_2/T + 10^6 C_3/T^2$

Phase A	Chemical Formula	$^1\alpha^{eq}$ 298.15 K	$^1\alpha^{eq}$ 573.15 K	$^1\alpha^{eq}$ 773.15 K	$^1\alpha^{eq}$ 1073.15 K	$^2C_1$	C_2	C_3	^{3}Range (K)	^{4}Reference
Quartz	SiO_2	1.0357*	1.0074	1.0027	1.0000*	−2.92	0	3.38	473–773	Clayton et al., 1972
Albite	$NaAlSi_3O_8$	1.0298*	1.0055*	1.0015	0.9991	−3.41	0	2.91	623–1073	O'Neil and Taylor, 1967
Anorthite	$CaAl_2Si_2O_8$	1.0206*	1.0027*	0.9998	0.9980*	−3.82	0	2.15	673–923	O'Neil and Taylor, 1967
Muscovite	$KAl_3Si_3O_{10}(OH)_2$	1.0231*	1.0034*	1.0001	0.9982*	−3.89	0	2.38	673–923	O'Neil and Taylor, 1969
Diopside	$CaMgSi_2O_6$		1.0002*	0.9989	0.9990	4.076	−9.176	4.00	673–1118	Fit to Matthews et al., 1983
Rutile	TiO_2			0.9941*	0.9974*	0.96	0	−4.10	848–1048	Addy and Garlick, 1974
Hematite	Fe_2O_3	1.0061				−12.3	0	1.63	273–393	Yapp, 1990
Magnetite	Fe_3O_4		0.9927*	0.9919*	0.9950	−3.70	0	−1.47	(773–1073)	Bottinga and Javoy, 1973
Scheelite	$CaWO_4$	1.0098*	0.9984	0.9965	0.9953*	−5.87	0	1.39	(373–823)	Wesolowski and Ohmoto, 1986
Uraninite	UO_2	1.0007	0.9923	0.9933	0.9952	4.42	−13.29	3.63	(273–1273)	Hattori and Halas, 1982
Calcite	$CaCO_3$	1.0288	1.0056	1.0017	0.9995*	−2.91	0	2.78	273–773	O'Neil et al., 1969
Strontianite	$SrCO_3$	1.0274	1.0049	1.0012	0.9991*	−3.26	0	2.69	273–773	O'Neil et al., 1969
Witherite	$BaCO_3$	1.0250	1.0036	1.0000	0.9980*	−4.25	0	2.57	273–773	O'Neil et al., 1969
Dolomite	$CaMg(CO_3)_2$	1.0351	1.0083	1.0038		−1.52	0	3.20	573–783	Northrop and Clayton, 1966
Siderite	$FeCO_3$	1.0322*	1.0060	1.0017*	1.0013*	−3.50	0	3.13	306–470	Carothers et al., 1988
Malachite	$CuCO_3Cu(OH)_2$	1.0331‡				2.66‡	0	2.66	273–323	Melchiorre et al., 1999
Barite	$BaSO_4$	1.0273*	1.0023	0.9982*		−6.81	0	3.00	453–623	See Friedman and O'Neil, 1977
Anhydrite	$CaSO_4$	1.0319*	1.0051	1.0007	0.9981*	−4.72	0	3.21	373–823	Chiba et al., 1981
Sulfate ion	SO_4^{2-}	1.0320*	1.0048	1.0003*		−5.12	0	3.251	346–621	Lloyd, 1968
H₂O vapor	H_2O	0.9907	0.9993*			2.0667	0.4156	−1.137	273–373	Majoube, 1971
CO₂ gas	CO_2	1.0412	1.0138*			−15.3	16.60	0	271–358	O'Neil and Adami, 1969

[1] Calculated value of α_{eq} at 25°C (298.15 K), 300°C (573.15 K), 500°C (773.15 K), and 800°C (1073.15 K), reflecting ambient, hydrothermal, subsolidus, and magmatic conditions. Extrapolated values are marked with an asterisk; no value is given if the extrapolation is highly inaccurate. ‡The determination for malachite assumes that the acid fractionation factor is identical to that for calcite.

[2] Additive terms may be corrected to reflect a uniform value of 1.0412 for the CO_2–H_2O fractionation factor at 25°C determined by O'Neil et al. (1975).

[3] Range of experimental determination to nearest degree in kelvins. Ranges for calculations are given in parentheses.

[4] Complete references for most sources are given in O'Neil, J.R. (1986) *Rev. Mineral.* **16**, 1–40; or Friedman and O'Neil (1977) *USGS Bull.* 440 K.K.

Index

244